The Endless Reconstruction and Modern Disasters

"Deftly interweaving theoretical, historical, spatial and social analysis, the authors illustrate powerfully (with global relevance) the irony of capitalist modernity embodied in the century-long national post-disaster reconstruction project in Messina: the creation through bureaucratic territorial "development" of the conditions of precarity and dependency for which those inhabiting and reworking for generations that "temporary" politically and physically constructed landscape are blamed."

—Ann Kingsolver, *Professor of Anthropology, University of Kentucky, USA*

"Disasters last a long time. The interminable reconstruction of Messina was precursory of the spatial and social configurations of inequality that still characterize the city. In 1908, Messina prefigured contemporary disaster relief. For its acute analysis, its compassionate ethnography, its theoretical skill in weaving space with social class and state with financial capitalism, Farinella and Saitta's book is essential reading."

—Magali Sarfatti Larson, *Professor Emeritus of Sociology at Temple University, USA*

"Domenica Farinella and Pietro Saitta have written a powerful analysis of the 1908 Messina earthquake that reveals its long-lasting impact on the destinies of the city and its people. They show how the earthquake remade Messina, as the fitful rebuilding both ensured the development of an impoverished working class and a bourgeoisie devoted to a rentier economy. Astonishingly clear, acutely written—an important contribution."

—Michael Blim, *Professor of Anthropology, The Graduate Center, City University of New York, USA*

Domenica Farinella • Pietro Saitta

The Endless Reconstruction and Modern Disasters

The Management of Urban Space Through an Earthquake – Messina, 1908–2018

palgrave
macmillan

Domenica Farinella
Dipartimento Scipog
University of Messina
Messina, Italy

Pietro Saitta
Dipartimento Cospecs
University of Messina
Messina, Italy

ISBN 978-3-030-19360-7 ISBN 978-3-030-19361-4 (eBook)
https://doi.org/10.1007/978-3-030-19361-4

This Palgrave Macmillan imprint is published by the registered company Springer Nature Switzerland AG
The registered company address is: Gewerbestrasse 11, 6330 Cham, Switzerland

Acknowledgments

We would like to thank Fillea-Cgil of Messina for their moral and material assistance in the conduction of some important segments of this work. In particular, we thank Daniele David and Domenico Carnevale who have wholeheartedly promoted a study on the condition of the working class in a depressed Sicilian city.

We are also extremely grateful for the support and the patience shown by our editors at Palgrave Macmillan: Kyra Saniewski, Madison Allums, and Rachel Daniel. We thank, in particular, Ann Kingsolver for her invaluable help and her reading of early versions of this manuscript as well as Deborah Kay and Donatella Sutera for their linguistic and editorial consultancy.

Last but not least, we thank Valeria Farinella for her graphic support.

Contents

List of Figures

List of Tables

CHAPTER 1

Introduction

1.1 Messina: A Modern Catastrophe[1]

Very early on the morning of December 28, 1908, the city of Messina was hit by a devastating earthquake, measuring magnitude seven on the Richter scale, and lasting for about thirty seconds. This proved to be one of the most destructive events ever to hit the Mediterranean area. About five or ten minutes after the earthquake, there was also a huge tidal wave. In Sicily and Calabria, in an area of 600 sq. km., between 70% and 100% of all buildings were destroyed. In total, the victims amounted to 85,000. For Messina, in particular, the outcome was apocalyptic. A few seconds after the quake, the city was a mountain of debris, and at least 65,000 people in the city alone were dead (the figures are still debated) (Guidoboni 2010).

Rescue teams were very slow to arrive and, once on the spot, ineffective. Indeed, the entire reconstruction process was very slow, or rather, characterized by different speeds. Bureaucratic mechanisms were quite fast in certain regards, and very slow in others. In general, however, physical reconstruction was unhurried (De Pasquale 2006).

The reason for our interest in this historic catastrophe lies in the fact that, in spite of the temporal distance, the Messina disaster was a very modern event. The newspapers set up an international network that covered the disaster for months. Donations came from every corner of the world. Political leaders as well as the royal family visited Messina. The royals, and the queen in particular, were portrayed at the site of the

D. Farinella, P. Saitta, *The Endless Reconstruction and Modern Disasters*,
https://doi.org/10.1007/978-3-030-19361-4_1

tragedy in an endeavor to help the victims. Special financial agencies were rapidly created, and the allocation of resources to small groups of companies and individuals followed, as part of a regime of exception and emergency. Above all, the Messina earthquake was the first massive disaster to hit Italy after the Unification. The event thus acquired symbolic and political meanings for the national government of the time. In spite of the many failures and flaws, the State envisaged the disaster as an opportunity to perform according to modern notions of efficiency and effectiveness (Dickie 2008). As we have already said, ad hoc agencies and new financial levers and tools were thus applied (loans, fiscal deductions, etc.). Consequently, the disaster set in motion financial speculative processes related to the land, and determined the emergence of a new social class of real estate owners—something that produced a structural and cultural shift from a bourgeoisie of traders and merchants to that of rentiers and speculators (Barone 1982).

One of the findings in the book is that the "economy of disaster" and the "shock economy" (Klein 2007) are not specific features of the present. On the contrary, the elements that characterize contemporary disaster-related speculative processes were largely active at the very beginning of the past century and helped the formation of the present. Thus, Messina appears as an ideal space in which to observe the very long-lasting effects of a certain "modern" way of managing disasters.

The Messina earthquake provides many familiar elements for today's readers, and we believe it is especially worth revisiting because, in our view, the consequences of this disaster are still being felt. A century after the earthquake, thousands of families continue to inhabit shanties made of wood and asbestos, without basic facilities. These shacks—that is, what remains of the original temporary housing received by the international donors and government in the aftermath of the earthquake[2]—are the most symbolic and tangible element of such continuity. Established in the aftermath of the earthquake, and replaced at times, they became part of the urban landscape and can still be considered "a necessary evil" for many poor families.

The main concern of this study, then, is with the living conditions of this population of the margins, the reproduction of their subordination, and their individual "policies of resistance" to the processes of marginalization that invested them over the course of a century.

The extension of the city is another element of continuity: almost 70 kilometers, or 43 miles, in length for less than 250,000 people. This is, at least initially, the outcome of the early post-disaster speculative processes based on land expropriation. Related to this, there are the elements that compose the local "geography of exclusion" (Sibley 1995). The direction of today's marginality follows almost the same lines drawn right after the earthquake. In fact, poverty is concentrated in areas that were sketched out in the second phase of the reconstruction, when the shacks were moved from the city center to different locations. These locations were the site for the development of public housing projects, which were interspersed with the shacks, and, in certain cases, residential buildings for the lower middle class. Still in the year 2000, there were an estimated 3100 shacks, with 12,000 people residing in them (figures that have only slightly decreased, according to recent data. See: Comune di Messina 2018). At the root of today's stratification were the so-called *ribaraccamenti*, or the relocation of the shacks and temporary accommodation to areas situated outside the emergency plans and the new urban development plan (Borzì Master Plan) (Noto 2008). In short, the present urban geographical and social periphery developed in the years that followed the earthquake, and have become spaces that in the local, intimate imaginary and culture recall the ideas of marginality, crime, poverty, and, later, urban decadence.

Work is a further element of continuity. The economic marginality of the city that followed the seismic event led to the development of the construction industry and the reduction of other sectors—with the noticeable exception of the public sector. For the mass of unskilled workers present in the city, construction was therefore the natural outlet. From a structural point of view, with a dramatic shift from the production of goods to the provision of services, early processes of tertiarization get to characterize the economic life of the city.

The weakness of the manufacturing fabric is counterweighted by the growth in construction and commerce, and by employment in the public sector (Farinella 2009). The extension of manual and temporary employment in the public sector and in the building industry has shaped altogether those forms of income support that are at the basis of political patronage (Gribaudi 1980; Chubb 1982) and the Southern European welfare model (Ferrera 1996, 2010; Esping-Andersen 1999). Both today and in the past, the main alternative for the members of this class is emigration.

With the end of important businesses (mostly active in food processing), the layoff of hundreds of workers, and an increasing business mortality rate, in the early 2000s the weak manufacturing industry of the city has entered a deep structural crisis. Within this context, the public and para-public sectors, subjected to liberalization, privatization, and austerity policies, ceased to be factors of stabilization. The crisis of these sectors, therefore, has determined subsequent crises. Many public agencies, for example, are experiencing a liquidity crisis, which has led to protests over non-payment of wages. What the territory is experiencing, thus, is a further process of socio-economic marginalization, characterized by growth in income polarization and "bad jobs". That is, jobs that are mostly manual, underpaid, and to varying degrees "irregular" (at least, with regard to the payment of wages, working hours, and services requested). Activities, however, that confirm the findings of many previous studies on the forms of complementarity and integration of such bad jobs with "regular work" (Saitta et al. 2013).

Yet, the population of the city amounts today to about 234,000—almost 10,000 people less than at the beginning of the current decade.

Although early traces of this tendency were already visible in the 1950s, this massive trend toward depopulation has been consistently recorded since the 1980s (David 2008) and indicates a weakness in the fabric of local production, which became more evident over the past few years. With the partial exception of the post-war years, the contemporary history of Messina appears characterized by the gradual disposal of productive activities. Consistent with a trend that preceded the earthquake, the retail trade, service sector, and public administration gradually become the main areas within the local economy (Chiara 2011). The 1991 census data shows that, since the 1980s, the secondary sector had experienced a significant decrease, although the construction industry made up around 41% of that business segment. In 2009, there was a further increase in the population: of the 41,400 people employed in the secondary sector, 23,745 were employed in the construction sector (Istat-Prefettura di Messina 2010). In 2016, however, data from the periodic National Builders Association's report (Ance 2016) shows that, within the framework of a strong regional recovery in the construction sector, the number of people employed in this industry decreased in Messina by 26% compared to a previous survey carried out two years earlier.

Data of this sort is probably affected by the strong rates of irregularity in that segment of the labor market. However, the loss of over 5000 jobs in the secondary sector—as shown by the 2001 census—and which continued over subsequent years—is mostly real, and should be properly judged in the light of the depopulation that characterizes demographic trends (David 2008, p. 43). In the same figures, it is also realistic to see a shift of a significant number of workers into informal employment. If we see this fall, on the one hand, as the "surrender" of some of the workers, which can be associated with the presence of significant migration flows toward the North, on the other hand, it appears to be a possible indicator of the increasing precariousness of employment relationships within the industry.

On the opposite side of the class lines, the formation of the bourgeoisie followed similar patterns. From 1909, in fact, the establishment of new channels for financing reconstruction and the creation of special agencies for the financial management of economic flow delivered the city into the hands of small interest groups, reconfiguring it both socially and physically—sanctioning the appearance of a new class of "rentiers" (Barone 1982). Since then, a handful of families—some of which came from the province and only emerged into the public arena after the earthquake—have handed down their economic, political, and social capital from generation to generation. Large portions of the lower sections of the same bourgeoisie—that is, the upper middle class of professionals and entrepreneurs—depend both on this handful of families and on the public administration (the State, the Council, the Region, etc.) in order to survive. This phase saw the rise of a number of phenomena that are relevant to the modern city: for example, the consolidation of a new urban "underclass"[3] and the emergence of new forms of spatial segregation; the considerable extension of the city as a result of land speculation processes encouraged by the financial mechanisms of reconstruction; and, finally, as we have already said, the polarization of power and influence.

In light of these elements, this book aims to reflect on the long-lasting nature of disasters, and their ability to shape the future. Over the next pages, we will see how many of the classic elements that lie at the basis of undesired events—such as vulnerability, class divides, isolation, and so on—were reproduced in an "intensified" way after the disaster took place, creating a sort of undeclared founding principle of the new community to come. Moreover, in consideration of the long period of time over which this history develops, this book is also about the stratification of styles of governance and the opposition tactics that actors, situated at the top and the bottom of the local social ladder, put in place at different stages of national history. Thus the history of Messina also becomes a way to discuss the history of Italy, or, at least, of that old "ball and chain", which is how the dominant national elites spoke of and considered the South (De Francesco 2012). A way, in other words, to show how disasters generate and mirror ideas, convictions, and prejudices that make up the national intimacy. The city, in fact, is part of a larger and depressed macro-area, called the *Mezzogiorno*, that is, the South of Italy. The reconstruction process can therefore be seen in relation to a larger history concerning the many economic interventions that sought to develop this macro-area after the Second World War. Far from achieving the declared goals, these interventions (the Cassa per il Mezzogiorno, the various development policies, etc.) became a tool in the hands of local and national politicians to create a complex economic model of "dependent development", based on the clientelistic exchange between votes and public housing, or unskilled jobs in the public sector, fueled by public expenditure and financed by means of public debt. From the post-war period to the early 1990s, this development model, which was based on the electoral mobilization of a vast marginal class in return for its "particularistic inclusion" in the lower segments of the middle class, obtained a high degree of consensus in the South of Italy, allowing many to achieve reasonable and widespread household stability (an important aspect of Mediterranean welfare regimes, as we have already noticed) (Trigilia 1992; Farinella 2005). In the case of Messina, this model has been adapted to the specific framework generated by the earthquake, so that at the center of this mobilization was found that marginal class of people who still lived in the shacks or in other precarious accommodation—which, according to the recent update by the local Council census, still number 9744 inhabitants and 2722 families (Comune di Messina 2018). Although these figures have decreased compared to the c.15,000 in the 1960s (De Pasquale et al. 1960), the situation is still far from being resolved. Furthermore, to this number, we should also add those living in public housing or in other illegal houses built on the hillsides and close to the city torrents. However, regardless of their living conditions, the vast majority of these people are also in precarious jobs with uncertain wages, and have a low standard of living—so that housing and job insecurity intersect in complementary ways.

In more detail, our research deals with the long-lasting social consequences of speculative approaches to the management of disasters reflected in the case of Messina, and shows the forms of marginalization of the subaltern population that started in the aftermath of a remote disaster and still affects its life conditions and modes of reproduction: the "underclass" is segregated in specific areas of the town, in specific types of houses, and in specific informal jobs. Confronted with such levels of structural inequality, the only alternative left to the subaltern classes is to continue to "exist/resist" within the territory. In this sense, the perpetuation of informal dynamics and spontaneous housing processes in the South of Italy does not indicate backwardness and a dependency culture, but "modernity". Thus, in a local context, the subaltern classes develop a series of arrangements, strategies, and existence/resistance resources, showing an impressive capability for enacting within their environment. For this reason, at least in our view, the southern city in general, and Messina in particular, is a key area for ethnographic observation, and one which highlights how formal and informal are a single social factor belonging to the current history of capitalism.

Yet, the historical process underlying the formation of the new urban space at the center of our study reveals how neoliberalism, a catch-all expression in many contemporary analyses, is not a modern way of managing the economy, the space itself, and controlling the social classes, but it is something with much older origins. In our case, in fact, the establishment of economic models based on profit and speculation, the massive erosion of the territory for construction purposes, and the exploitation of an unskilled workforce forced to live in poverty seem to be the results of a process that began in 1908, during the reconstruction that followed the tragic earthquake that hit the city at the end of that year. A tendency that continued both during the Fascist era and later, with the so-called *sacco* (the plunder) of the southern cities in the years that followed the Italian economic "miracle" (1960s–1970s). A trend, finally, that was perpetuated by the crisis in the welfare state. The issue of the shanty-dwellers lies within this chronic framework of decline in the local economy (Chiara 2011). The perspective presented here aims also to emphasize the fact that, although parts of the South are seemingly going in different directions in comparison to other parts of the country, this is, more than the result of internal features, the outcome of complex forms of convergence between "core" and "peripheral" elites—a convergence of interests aimed at specializing the interior territories and assigning precise functions within the domestic division of labor. Unlike prevalent visions of the "Southern Question" in terms of development and underdevelopment, the eternal precariousness of the South is understood here, in particular, as a signal for the end of the welfare state and Fordism, and the advent of a late liberalism (or, if you prefer, neoliberalism), globally characterized by the same inequalities and forms of instability that have historically identified this part of Italy. A local contribution, in short, to the notion of "Brazilianization of the West" (Beck 2000, p. 1, passim).

1.2 The Organization of the Book

Chapter 2 introduces the notion of disaster and provides a selection of key themes derived from the current sociological and anthropological debate on undesired events. Disasters are seen as events that take place within existing class divides, and affect social groups in different ways. Such events are, so to say, "augmentors" of reality: that is, instances that accelerate trends and tendencies active in a society at a given moment. For example, as far as the institutions are concerned, the ideologies and regimes active at a given moment will affect the responses to a disaster and the solutions provided. In other words, similar disasters that took place under Roosevelt's *New Deal* and under Bush's neoliberal state—for example, Katrina—might not have generated the same responses due to the different understanding of the role of State (Powers 2006).

The way disasters augment reality includes the visibility of social structures and tendencies. In 1908 Messina presented a strong but declining economy. The earthquake, thus, has only accelerated the decline of the city. As a consequence, even the narrations, the self-representations, and symbolisms related to this calamity are involved. If other important disasters, such as the Great Chicago Fire or the 1906 San Francisco earthquake, are related to an internal narration of rebirth (Sawislak 1995; Dyl 2017), the Messina earthquake, instead, evokes the idea of an "end".

Likewise, disasters make the relations between the places affected and the political powers tangible. The response of the State and the attention reserved for places hit by a disaster mirror the distance, or the proximity, between these areas and the central power. In this respect, disasters are also measures of the importance of a place, that is, they make visible both the centrality of the ruling classes and the positions within the internal hierarchies of a nation (Wisner et al. 2004).

Finally, disasters—and this has certainly been the case with Messina—can be socially divisive events. Although this is not compulsory (Binu et al. 2008), nonetheless, calamities are often situations that feed social division and threaten social cohesion for a long time. Our study, then, deals with the process of the formation of distinct inner circles and divisions that are the very basis of the current social order, well over a century after the earthquake.

Chapter 3 deals with the economic, ideological, and intellectual framework within which both the disaster and the long recovery process took place. Messina and the South of Italy are seen here as places that occupy specific roles in the national division of labor since the Unity. These roles are related to the formation of markets, industries, and the labor force, formation processes that underwent important changes over the course of the decades and that were dependent on the international conjuncture, the technical changes, and the position occupied by the country in each season. Such structural changes were accompanied by historical analyses, public discourses, and the engagement of intellectuals. Nevertheless, there have always been elements of continuity in these types of discourse and in the real effects they produced. The goal of this chapter, then, consists of tracing both the changes and the elements of continuity, from the liberal policies and governments of the nineteenth century to Fascism, to the post-war Republic and neoliberal Italy of today. Suspended between sociology, anthropology, and economic history, this part of the book will help frame the Messina earthquake within a set of concepts, relationships, and functions that will hopefully broaden the vision of disasters as events concentrated in time—an attempt, in other words, to subtract the analysis from the limits of an investigation that focuses on "the short time span, for the individual and the event" (Braudel 2009, p. 173).

Chapter 4 investigates the polarized social order that has emerged in the city of Messina over a long period of time: starting with an analysis of the socio-economic conditions of the city prior to the earthquake, which highlights the progressive marginalization of the city before the event. This chapter reflects on the effects of the disastrous event of 1908, passing through the reconstruction phase of fascist policies that produced the "slumization" of the city, and the "Fordist" building policies of the second post-war era. Finally, the analysis explores the present neoliberal policies.

The chapter also shows how different conceptualizations of public space in the city of Messina, from 1908 onward, equally implied the confinement of marginal classes to specific areas of the city, within specific types of accommodation, in order to favor speculative operations by large landowners, to increase the ground rent and, finally, in order to build patronage networks aimed at vote buying.

From the early Fascist reconstruction to the present, endless policies were applied, the outcome of which consisted in the forced displacement of the population to the most peripheral and deprived areas of the city. The lack of services and green spaces in these marginal areas is associated with this "ideology of separation" from the rest of the city. An institutional determination, in other words, to establish internal, symbolic borders for the lower-income and disenfranchised population of the place.

This chapter describes also the role of public and private building during the generation of the capitalist development model, which puts both land exploitation and speculative income at the center of its frame. It was a model able to adjust and match both the post-war Keynesian-Fordist and contemporary contexts.

Finally, Chap. 4 is an inquiry into the historical continuity of class formation in a city. This part shows how the local development model is related to the socio-spatial hierarchization of the population, which confines the lower classes in a precarious job market and marginal neighborhoods placed at the edges of the city.

Such hierarchization determines a social order that hinges essentially on the division between a relatively small and "tertiarized" middle class (i.e., active in the services sector), and a much wider class of individuals, variously occupying subordinate positions. Some of these latter, in fact, are at the center of the exploration contained in Chap. 5.

Chapter 5 contains the findings of different research projects carried out between 2001 and mid-2017. During these years of intense research activity conducted by the authors, it was possible to analyze the city from different standpoints (i.e., from the position of the precarious workers in the construction sector and public sectors, as well as from the perspectives of the slum-dwellers). Eighty-five in-depth interviews with precarious and unskilled workers in the public hospital and building sector, and different periods of participant observation were carried out in different popular suburbs in the southern part of the city (in particular, Santa Lucia-San Filippo, Rione Gazzi,) also focusing attention on the adjacent areas, where many residents live in social housing and shanties (this part of the research is also presented in the Chap. 6). Moreover, a long period of observation was conducted in the headquarters of a major trade union (Fillea-Cgil). At the center of this latter portion of the study were the construction workers represented by the union, their political orientations, and lifestyles. Finally, in 2017 one of the authors joined a collective involved in the occupation of public buildings in order to fulfill the need for housing for evicted families. Observations were also carried out in other popular districts, in order to study the territory in greater depth. This joint research aimed to explore the ways in which subjects belonging to subaltern classes have made use of the gaps left open by local powers, in order to access forms of social mobility, even if they are fragmented and unstable.

This chapter focuses specifically on those "working poor" who make up much of the present labor force, in the city as well as internationally (Munger 2002). Through biographical analyses, the chapter provides some reflections on the subaltern condition in the city of Messina and, to some extent, in southern Italy as a whole. We argue that the existential paths and the connected visions of the studied population should be seen as intimately political processes—aimed at moderating asymmetries and structural injustices—as well as the outcome of emotional processes characterized by the systematic frustration of aspirations and desires.

Chapter 6 focuses on the popular suburbs. As we have shown in Chap. 4, the results of the construction of urban public space in Messina relegated marginalized subjects to specific areas of the city, housing them in specific types of accommodation, with the effect of generating a speculative economy that was linked to the patronage system. These endless policies have led to the forced displacement of lower-income citizens to the most peripheral quarters, in which the lack of services is associated with symbolic and material separation from the rest of the city.

Directed toward the opportunistic interests of a specific powerful social class, public space has become an increasingly "denied" space, where power relations linking patrons to clients are crystallized—by transforming a "right" into a "favor". It is a process of marginalization created as a result of public policies, but in common discourse it is embodied in the unacceptable formula of a "shanty culture", where the individuals living in these houses inherit an immobility and an inability to think differently, to the point of reproducing this sort of logic even in public housing. In the face of this sort of symbolic violence, reaffirmed by the way urban space is organized, the individuals put specific social mobility strategies into practice geared toward becoming homeowners, so they are able to imitate the middle-class lifestyles to which they so strongly aspire, but which are in fact denied them. This part of the book investigates the trajectories and the patterns of segregation faced by this population. We show how their untidy, ugly, and shameless presence within the territory expresses their practices of "resistance" toward power itself, and determines a completely new configuration of urban space leading to a different rationale from that of formal institutional regulations.

Chapter 7 shows some of the representations concerning the slums and slum-dwellers, showing how these often stereotypical representations are periodically politically mobilized by the various actors in the field (journalists, politicians, the slum-dwellers themselves, and the local population), in order to achieve strategic objectives. More specifically, we can see how journalistic narratives "subalternize"[4] the slum-dwellers, through the emphatic but fictitious connection between the earthquake, poverty, and the shacks, in which the shack-dwellers are seen as victims of a system of patronage, exploiting them with false promises in exchange for their votes, but it also shows how they are victims of themselves, unable to abandon the shacks and a culture of dependence. In our view, rather, the slum-dwellers use of this narrative register does not express their true cultural adhesion, but it appears to be a tactical choice that can help them to legitimize their personal and "political" claims (to obtain public housing), and which therefore needs to be carefully "staged" in certain situations, and avoided in others. Once again, the slum-dwellers appear to be active subjects capable of overturning their subaltern position and of producing "political" tactics, aimed at obtaining certain effects and at contrasting others. Finally, we show how politicians (both on a national level and on a local level) strategically exploit the symbolic field implicit in the stereotypical representations of the slum-dwellers, in order to strengthen their own political position—this being what we can call an endless series of overlapping tactic and strategic manipulations.

The final chapter (Chap. 8) discusses the end of the "model" presented over the course of the book. This type of model, based on the use of public resources and patronage, came to an end at the beginning of this decade, when the construction industry, which has traditionally been the space of inclusion for the local labor reserve army, came to an abrupt halt. Due to the lack of resources generated by austerity policies and the new concerns with the erosion of the territory—the decrease in public works and private building in recent years has left many unemployed. The city's public sector is also seeing cutbacks. Public authorities, especially military infrastructure (hospitals or headquarters), have been moved to nearby cities (Catania, primarily). Subsequently, the great 2008 recession fueled massive emigration processes. Since the beginning of the crisis, in fact, the city has lost about 3000 people a year. In addition, the city came extremely close to bankruptcy and, for much of the old political class, the legal threat to become ineligible for public office for a decade almost became a reality. For the local clientelistic networks, it became impossible to use public resources to feed their connections. Consequently, in 2013 the political landscape changed dramatically. Renato Accorinti, a pacifist, "anarchist", and an activist from the social movement sphere, became mayor of the city. His electoral base was broad and included the middle classes as well as the inhabitants of the city's slums—all of them hailing the promised "revolution". If the expectations were, as some local observers noticed, "messianic", the results were disappointing (Palumbo 2016). The new local government avoided bankruptcy, but it embraced, at least partially, the austerity measures imposed by the national government. Subsequently, the large movement that had supported the mayor's split and the people from the poorer areas became nostalgic, as the regional elections at the end of 2017 demonstrated. Over the course of these elections, in fact, the classic political forces that ruled the city until the 2013 defeat came back stronger than ever (in spite of the many opposition movements that have consolidated their position in Italy and on the rest of the island during the past few years, especially the Five Star Movement). Only a few months later, in June 2018, Renato Accorinti was soundly defeated in the city elections. The new Mayor Cateno De Luca, however, is not part of the traditional elites, but a classic "populist". De Luca is a long-standing professional politician who, nevertheless, presents himself as an "outsider" external to any circle of power. In spite of the structural nostalgia still present in the city, the long post-earthquake era is over and a new phase—perhaps even unhappier than the previous one—is commencing.

Notes

1. This book is the result of joint contributions by the authors. Each concept, page, section, and note is the fruit of this common work. However, for academic reasons and due to Italian legislation, the even-numbered sections should be attributed to Domenica Farinella and the odd-numbered ones to Pietro Saitta.
2. Indeed, the Messina disaster was also "modern" in the way that it was able to generate international aid organizations and donations. Although no study has analyzed this mechanism as a "system"—in the same way that disaster research has with cases such as Haiti (Schuller 2012)—and the available studies focus only on specific donors and nations, we can nonetheless draw an idea of the extension and logic of this organization of aid on the

basis of the contributions by: Di Giacomo (2008), Mazzoli (2008), Di Paola (2010), Andreides (2010), Caroniti (2010), Bottaro (2010).

3. The term "underclass" is highly problematic, and it expresses an ambiguous and controversial concept (Bhalla and Lapeyre 1999, p. 96). We agree with Katz (1989, p. 10) who has noticed that this word is a synonym for the undeserving poor. Although we continue to use it from time to time, we are aware that the experience of this societal component stems from processes that affect the whole of society (Buck 1996, pp. 278–279). Our use, then, is a synonym for "lumpen", or the lower levels of the proletariat—the outcome of a structural order.
4. Subalternity is a Gramscian concept whose interpretations and meanings depend on different intellectual traditions developed in different continents and political milieus. By utilizing this word, we point at the hierarchies within society and at those processes, led by local elites, which produce oppressed groups within a nation (in terms of wealth, ethnicity, cultural capital, etc.). Groups, moreover, whose (undesired) existence allows the self-definition of the dominant components of a nation. With regard to this latter and perhaps obscure aspect, the concept of subalternity expresses the exposition of powerless groups to public narrations, definitions, and stigmas produced by those who are in a position of power. Not by chance, in Gramsci's terms, "For a social elite, the members of subaltern groups always have something of a barbaric or pathological nature about them" (Gramsci 1996, Q3 §12).

References

Ance. (2016). *Rapporto congiunturale sull'industria delle costruzioni in Sicilia*. Palermo: Ance.

Andreides, G. (2010). Il terremoto di Messina e l'Ungheria. In A. Baglio & S. Bottari (Eds.), *Messina dalla vigilia del terremoto del 1908 all'avvio della ricostruzione* (pp. 321–330). Messina: Istituto di studi storici Gaetano Salvemini.

Barone, G. (1982). Sull'uso capitalistico del terremoto. Blocco urbano e ricostruzione edilizia a Messina durante il fascismo. *Storia Urbana, 10*, 47–104.

Beck, U. (2000). *The Brave New World of Work*. Cambridge: Polity Press.

Bhalla, A. S., & Lapeyre, F. (1999). *Poverty and Exclusion in a Globalized World*. New York: St. Martin's Press.

Binu, J., Mawson, A. R., Payton, M., & Guignard, J. C. (2008). Disaster Mythology and Fact: Hurricane Katrina and Social Attachment. *Public Health Report, 123*(5), 555–566.

Bottaro, G. (2010). Theodore Roosevelt, gli Stati Uniti e il terremoto di Messina del 1908. In A. Baglio & S. Bottari (Eds.), *Messina dalla vigilia del terremoto del 1908 all'avvio della ricostruzione* (pp. 365–378). Messina: Istituto di studi storici Gaetano Salvemini.

Braudel, F. (2009). History and the Social Sciences. The Long Durée. *Review, 32*(2), 171–203.

Buck, N. (1996). Social and Economic Change in Contemporary Britain: The Emergence of an Urban Underclass. In E. Mingione (Ed.), *Urban Poverty and the Underclass: A Reader* (pp. 278–298). Oxford: Blackwell.

Caroniti, D. (2010). Michelopoli. In A. Baglio & S. Bottari (Eds.), *Messina dalla vigilia del terremoto del 1908 all'avvio della ricostruzione* (pp. 331–338). Messina: Istituto di studi storici Gaetano Salvemini.

Chiara, L. (2011). *La modernizzazione senza sviluppo. Messina a cento anni dal terremoto.* Firenze: Le Lettere.

Chubb, J. (1982). *Patronage, Power, and Poverty in Southern Italy.* Cambridge: Cambridge University Press.

Comune di Messina - Dipartimento Politiche per la casa. (2018, August 23). *Relazione al Sindaco sullo stato del risanamento.* Messina: Comune di Messina.

David, P. (2008). Territorio e demografia. In M. Centorrino & P. David (Eds.), *Senza corona e senza scorta. Quarto rapporto sull'economia della Provincia di Messina* (pp. 38–46). Messina: Armando Siciliano Editore.

De Francesco, A. (2012). *La palla al piede. Una storia del pregiudizio antimeridionale.* Milano: Feltrinelli.

De Pasquale, D. (2006). *Le mani su Messina. Prima e dopo il terremoto del 1908.* Messina: Armando Siciliano Editore.

De Pasquale, L. C., Pezzino, G., & Vincenzo, F. (1960, April 6). *Proposta di Legge "Provvedimenti per il risanamento e lo sbaraccamento della città di Messina".* Rome: Camera dei Deputati.

Di Giacomo, S. (2008). Il terremoto di Messina del 1908 e gli aiuti tedeschi. *Storia e Futuro, 17*, 1–18.

Di Paola, M. T. (2010). L'intervento britannico nel terremoto del 1908. In A. Baglio & S. Bottari (Eds.), *Messina dalla vigilia del terremoto del 1908 all'avvio della ricostruzione* (pp. 297–320). Messina: Istituto di studi storici Gaetano Salvemini.

Dickie, J. (2008). *Una catastrofe patriottica. 1908: Il terremoto di Messina.* Rome and Bari: Laterza.

Dyl, J. L. (2017). *Seismic City. An Environmental History of San Francisco's 1906 Earthquake.* Seattle: University of Washington Press.

Esping-Andersen, G. (1999). *Social Foundations of Postindustrial Economies.* Oxford and New York: Oxford University Press.

Farinella, D. (2005). *Privatizzazione e cambiamento nel servizio pubblico*. Milano: FrancoAngeli.
Farinella, D. (2009). Messina, un'economia in transizione, tra cambiamento e marginalità. In M. Centorrino & P. David (Eds.), *Le città della Fata Morgana* (pp. 47–84). Milano: FrancoAngeli.
Ferrera, M. (1996). The 'Southern Model' of Welfare in Social Europe. *Journal of European Social Policy, 6*(1), 17–37.
Ferrera, M. (2010). The South European Countries. In F. G. Castel, S. Leibfried, J. Lewis, & C. Pierson (Eds.), *The Oxford Handbook of the Welfare State* (pp. 616–629). Oxford: Oxford University Press.
Gramsci, A. (1996). *Prison Notebooks Volume II*. New York: Columbia University Press.
Gribaudi, G. (1980). *Mediatori. Antropologia del potere democristiano nel Mezzogiorno*. Torino: Rosenberg & Sellier.
Guidoboni, E. (2010). Il terremoto e il maremoto del 1908: come nasce una catastrofe. In A. Baglio & S. Bottari (Eds.), *Messina dalla vigilia del terremoto del 1908 all'avvio della ricostruzione* (pp. 149–174). Messina: Istituto di studi storici Gaetano Salvemini.
Istat-Prefettura di Messina. (2010). *Messina in cifre 2010*. Messina: Prefettura di Messina.
Katz, M. B. (1989). *The Undeserving Poor. From the War on Poverty to the War on Welfare*. New York: Pantheon Books.
Klein, N. (2007). *The Shock Doctrine. The Rise of Disaster Capitalism*. New York: Henry Holt and Company.
Mazzoli, E. (2008). *"Nei giorni di tanta incommensurabile sciagura…". Trieste, l'impero e il terremoto di Messina del 1908*. Milano: Biblion.
Munger, F. (Ed.). (2002). *Laboring Below the Line. The New Ethnography of Poverty, Low-Wage Work, and Survival in the Global Economy*. New York: Russell Sage Foundation.
Noto, A. G. (2008). *Messina 1908. I disastri e la percezione del terrore nell'evento terremoto*. Soveria Mannelli: Rubbettino.
Palumbo, B. (2016). Debt, Hegemony and Heterochrony in a Sicilian City. *History and Anthropology, 27*, 93–106.
Powers, M. P. (2006). A Matter of Choice: Historical Lessons for Disaster Recovery. In C. Hartman & G. D. Squires (Eds.), *There Is No Such Thing as a Natural Disaster. Race, Class and Hurricane Katrina* (pp. 11–34). New York: Routledge.
Saitta, P., Shapland, J., & Verhage, A. (Eds.). (2013). *Getting by or Getting Rich? The Formal, Informal and Criminal Economy in a Globalised World*. The Hague: Eleven Publishers.
Sawislak, K. (1995). *Smoldering City. Chicagoans and the Great Fire. 1871–1874*. Chicago: The University of Chicago Press.

Schuller, M. (2012). *Killing with Kindness. Haiti, International Aid, and NGOs.* New Brunswick, NJ: Rutgers University Press.
Sibley, D. (1995). *Geographies of Exclusion. Society and Difference in the West.* London and New York: Routledge.
Trigilia, C. (1992). *Sviluppo senza autonomia. Effetti perversi delle politiche del mezzogiorno.* Bologna: Il Mulino.
Wisner, B., Blaickie, P., Cannon, T., & Davis, I. (2004). *At Risk. Natural Hazards, People's Vulnerability and Disasters.* London and New York: Routledge.

CHAPTER 2

Researching Disasters: Theories for a Case Study

2.1 The End of the World

In a thorough study on the imageries generated by disasters, Noto (2008, p. 18) observes that the distinction between disasters and catastrophes lies in the reversibility of the undesired events. The former term indicates an event whose consequences can be reabsorbed over a short lapse of time; the latter, instead, points to a sudden crisis the effects of which cannot be reversed. To put it in the words of the seminal Italian anthropologist Ernesto De Martino (2002), in sum, catastrophes allude to the end of the world, or to "cultural apocalypses", which may, or may not, contemplate the idea of a rebirth.

These annotations aptly introduce some of the issues that the 1908 Messina earthquake saw at work in the aftermath of the event and over the following decades. One of key questions of this study consists in establishing whether that occurrence has been a disaster or a catastrophe, and whether a rebirth followed or not.

This chapter introduces the reader to a disaster that took place in the Sicilian town of Messina in 1908, and it provides a selection of key themes derived from the current sociological and anthropological debate on undesired events. Particular elements that pertain to local history and general findings highlighted by the international literature on disasters are put in connection in order to show the elements of continuity between heuristic levels, and to identify a set of concepts that will lead us through the entire study.

D. Farinella, P. Saitta, *The Endless Reconstruction and Modern Disasters*,
https://doi.org/10.1007/978-3-030-19361-4_2

In particular, we see disasters as elements that can revitalize old fractures and lines of separation that are at the very basis of the processes of national formation. Calamities expose and make tangible different positions and perceptions related to time, space, and economy. Thus, from our professional perspective, beyond being tragic phenomena, disasters are cultural events disguised as politics. That is, a time of emergency that wipes out institutional checks and balances, or any other similar superstructure, and makes visible latent forces and different ways to cope with the unexpected and undesired.

But disasters are also a way to produce new orders and new understandings of the situation. Disasters, and post-disaster phases in particular, are modes to produce novel forms of stability and configuration of social relations that reflect the public culture of a given time. The concept of public, however, implies also the notion of plurality: the same plurality at the very basis of societies segmented by class, power, education, and opportunities. Disasters, thus, provide the observer with the chance of exploring different public cultures within a society and, above all, the "class conflict" that takes place in the moment that an order is to be established.

In accordance with this premise, we believe that the seismic event that destroyed the Sicilian city at the very beginning of the past century marked a divide for the city, which, more than a century later, appears still suspended between feelings of nostalgia and a sense of loss. Countless local intellectuals, websites, and Facebook pages still celebrate the city that disappeared at the dawn of a tragic morning more than a century earlier. The beauty of that city, its monuments, churches, and streets are still at the center of reenactments and debates that evoke a "structural nostalgia", that is, in Herzfeld's words, the "collective representation of an edenic order—a time before time—in which the balanced perfection of social relations has not yet suffered the decay that affects everything human" (Herzfeld 2005, p. 147). Certainly, some corrections should be added to this definition, as the lost time to which local forms of nostalgia refer to is not unknown. Moreover, the sense of decay mentioned by Herzfeld has, in this case, very specific objects. That is, the ideas of beauty, wealth, commercial power, and an entrepreneurial spirit. Perceptions concerning space, materiality, and moral virtues of the inhabitants are, therefore, at the center of this discourse. In this respect, it is instructive that a paper presented by a group of natural scientists at a conference on gas geochemistry in 2007 was successful and circulated very widely. The study, which has never been published on a scientific journal, suggests that prior, during, and after the 1908 earthquake, massive emissions of gas radon could have altered the DNA of the inhabitants of the city of Messina (Italiano et al. 2008). Several local newspapers and magazines reported the news, and they suggested that such alterations might have caused changes in attitude and behavior of the local population. Until recently, one such magazine, for instance, claimed that "to accompany this hypothesis to the actual political and economic fracture that followed the earthquake, is something that could have much to tell us on a population that, today, is considered 'babba' [stupid] and sluggish" (Redazione 2016). Following popular beliefs and simplifications on the relationship between genetics and individual attitudes, this article, interestingly, connects the scientific "fact", related to possible alterations in the DNA of the population, to the supposed "stupidity" of the city—the *città babba*. In Sicily, *babbe* were considered a handful of cities, including Messina and Syracuse, that were said not be home to mafia families. But leaving temporarily aside this important aspect, which hides a clear ambivalence toward organized crime, it is significant that *Messinesi* depict themselves, and are depicted from outside, as stupid. Yet, an associated and popular term used to define the psychology of the city is *buddace*, a word which refers to a thin fish with a large mouth, the sea bass. Loudmouth or blowhard, thus, are possible translations for this expression, which indicates a collective personality trait characterized by the tendency to bragging and boasting—or saying without doing.

To accompany this sense of collective ineptness, there are common complaints about the role that the nearby city of Catania plays in diminishing the importance of Messina at both an economic and institutional level. More than just a mere competitor, Catania is commonly perceived as that evil entity that diminishes the institutional and economic importance of Messina. For instance, Catania is said to "steal"—that is, to subtract—important institutional organs and institutions of control and regulation such as the port authority, the military command, and so on. Without taking into account that Catania is home to many important local economic initiatives, and that it hosts multinational brands such as Ikea, and a few international companies operating in the electronics sector within the so-called Etna Valley. Even thieves, robbers, and criminals active in the city of Messina are said to have come from Catania.

This way of (self-)describing the situation of Messina in relation to negative collective traits and to its most important economic competitor is something that emerges after the 1908 earthquake. Until that moment, in fact, the city excelled both in comparison to Catania and the other big Italian cities. At the turn of the century, in the 1890s, the Port of Messina, for example, had been the first of the island and the sixth of the nation. Due to structural changes in the national and international logistics system, and the overall economy of the island, and thanks to the emergence of new southern ports, at the time of the 1908 earthquake, Messina was still the third port of the island and the tenth of the country. Catania was, so to say, still off the radar. In the years between 1908 and 1911, right after the earthquake, the exchanges that took place in the port of Messina decreased by 16%, while the traffic in the port of Catania increased by 18% (Battaglia 2010, p. 68). Since then, as is shown in Chaps. 3 and 4, the economic and political marginalization of Messina in the regional and national scenario increased significantly.

The sense of referring to these elements that shape both the cultural intimacy of this place and its economic situation lies in the idea that there are undesired events—namely catastrophes—which last over the decades and the centuries, and determine the modes of representation of places. Oliver-Smith (2002, p. 25) observed that disasters generate multiple interpretations of the event and the related processes, and they are experienced differently by different groups and individuals. While this is true and it also entails the idea that undesired events are social constructions whose meaning and effects depend, among other things, on the position of the individual and collective actors in the social ladder, such events are also cultural elements. That is, elements that generate collective reactions and, thus, compose the common sense of a place and its "structural amnesias" and "memories". In her discussion of the tremendous environmental and social issues that characterize the industrial areas of today's Louisiana, Hochschild (2016, p. 51) notices, for example, that "the general talk around town was that the choice was between the environment and jobs", and that, as a consequence, "one shouldn't get to hung up on the environment". So, Hochschild argues, the memory of Southern environmental glory fell. Borrowing the concept from Evans-Pritchard, who observed how the pastoral people of Sudan called the Nuer had a prodigious memory for some things and forgot about others, Hochschild recuperates also this notion of "structural amnesia", that is, the selective process that pushes people to remember or forget things in scenarios of disaster as well as in normal times.

If in Hochschild's case, the paradox is that the good things of the past are forgotten in favor of the "bad" things of today (whereas the resignification process is such that in Louisiana, pollution and disease evoke also the desirable ideas of wealth and economic well-being), in the case of Messina the earthquake is, perhaps more comprehensibly, perceived as the event that caused the end of a world. In fact, the seismic event was not chosen, and most of the people could not draw any advantage from it—although some certainly did.

Thus, not only "good things" are remembered, but the bad ones can also be utilized to interpreting reality and making sense of it. However, one should keep in mind that these selected memories do not attain only to the public sphere. These are not only a type of popular or oral historiography that reconstructs "from below" reasons and motives behind present realities. Disasters, and above all disasters that turn out to be catastrophes, mark forever the end of worlds, and so shape individual and collective sentiments. They are events through which affects are produced and, perhaps, also governed. In his description of the reconstruction processes that followed different disasters in Honduras, Mexico, and the USA, Barrios (2017) shows how disasters shape people's intimate feelings, and how such feelings produce public, if not political, effects. In other words, effects determine affects—that is, modes of perceiving reality do not emerge as emotions, nor as passions but as interior feelings that cannot easily be rationalized and expressed into words, and that produce membership or self-exclusion, sense of loss, or satisfaction. Affects, moreover, are shared and become, thus, public and implicitly political.

Messina's post-earthquake experience shows how such a worldview, once it is structured and rooted, becomes a cultural trait. That is, a narration and a way of experiencing reality that is repeated and transmitted from generation to generation. The narration of a time that has gone forever.

2.2 Civility in the Post-disaster City

The passage of time and the encounters with "modernity" do not change this worldview; on the contrary, such encounters show the relevance of the "complex of being 'backward'" (Saugeres 2002, p. 146). In the aftermath of the disaster, when the city was still under the debris, influent national and international witnesses, such as journalists, medical doctors, and army officers who reached Messina from the north of Italy or from abroad, to describe the situation or to help, wrote a number of reports that systemically denigrated the city. Noto (2008, pp. 294–299) has collected many of these reports from the field. In general, these witnesses complained that the survivors did not support the rescue operations, and that they appeared idle and indifferent. In the influential cultural review *Nuova Antologia*, an editorial article stated that the disaster involved many peasants for whom the death of a relative was tantamount to that of a pet, and that this population needed "a valid moral education by the Italian State" (Noto 2008, p. 295). Likewise, the socialist intellectual Tommaso Rossi-Doria claimed that the "lack of heart and civicness" of the citizens of Messina was due to the "millenniums of abandonment" by the State that characterized the history of their city (Noto 2008, p. 295). And, in general, the numbers of articles that expressed similar opinions were countless.

While these discourses reflect a positivist view of the South which was very common during those years and that lies at the heart of that "Southern Question" analyzed, among others, by Schneider (1998) and Moe (2002), it is interesting to observe how the theme of civilization is often repeated in current public discourse (see Chap. 3 for a detailed discussion). An analysis of such a discourse would show, in fact, that Messina is a "dual city" that sees the contrast of two classes: the "civilized" and their "other", the *zalli*—that is, the boors.[1] As is shown in Chap. 6, today's Messina, still a city of shacks, slums, and dilapidated housing projects, has a problem with its "underclass". Petty traders, street vendors, little vandalisms, and double parking are some of those elements and presences, common to most large- and medium-sized cities in the world, that, for most commentators, mark the impossibility of pursuing civility in the city. While most cities around the world deal with these forces and pursue "zero tolerance" policies (Roy and Alsayyad 2004), the relevant element is that the local dominant discourse spins around the positivist themes we saw at work at the beginning of the past century. In a photobook on the shacks of Messina, for example, a local journalist argued that those who live in these spaces inherit "an immobility and an inability to think differently" (Pruiti 2010, p. 33). "Uncivilized" is one of the words most commonly thrown around by the local media and the common people belonging to the middle class. Not by chance, in 2015, the debate on the limits imposed on the traffic in the coastal neighborhood of Faro over the summer, has produced a public debate mostly centered on the notion of civility and the comparison with similar regulations in the "more civilized cities". And this is only one of the many available examples of local public debate that show the persistence of such longing for a certain idea of (missed) modernity.

Clearly, civility is not a theme specific to Messina, and it is actually one of the key issues of modernity. From Italy to France, and from the USA to the African and Asian colonies, virtually, no country has escaped the legacy of positivism and "progressivism" (Hofstadter 1955; Rabinow 1995; Gibson 2002). Nevertheless, ways of applying these ideologies have differed from country to country, and the legacy of such modes of application took local forms. Their specific and local meanings, in other words, are subjected to local history. In this respect, it is interesting to observe that it is only after the 1908 earthquake that a city that dominated a large island, and was home to an international bourgeoisie whose exchanges included the other Mediterranean countries as well as the northern European states and the USA (Battaglia 2003), discovered that it was indeed a "city of peasants" posed at the borders of civilization. There is, therefore, a curious resonance between this new way of being reimagined and narrated in the public space in the aftermath of the big earthquake, and the language of the contemporary upper class and its intellectuals. Something, indeed, that resembles the new terms, complexities, and aspirations of postcolonial spaces such as the Calcutta narrated by Roy—that is, a "gentleman city" which aims to reproduce a "bourgeois public sphere", shaped by distinctive notions of social order and morality borrowed by the language and ethos of the former colonizers (Roy 2004, p. 161).

Of course, in the case of Italy, the language of civility is not specific to Messina. And even in the northern cities, this theme is still very common, as a search of the local editions of newspapers would show (see, e.g., the website "www.romafaschifo.com"). Nevertheless, in places like Milan, Turin, and the many other centers that after the Second World War were home to flows of workers from the Southern regions of the country and from abroad, the issue of civility acquires different meanings. Here, the local social question mirrors the Southern Question. As the works of Montaldi and Alasia (1960), Fofi (1964), or Foot (2001) have shown, in the modern age, the uncivilized class in these hegemonic cities was represented not as much by the local underclasses but by the southern immigrants—that is, an external "other". Since the Unity, in fact, the dominant discourse opposes the hardworking northerners and the idle southerners (Petraccone 2000). And despite the nationalist claims and attempt to produce a unitary representation of the country, Italy has always been intimately divided by ethnic and almost racial lines, as the lengthy Northern League's aspiration to secession has shown (Huysseune 2006, p. 207 passim). Moreover, such representations do not belong only to a tiny but influential political elite that aims at pursuing a particular project. On the contrary, they are rooted in the internal culture of Italy and slow to die. If visible expressions of internal racism directed against the southerners, now turned into institutional xenophobia, were part of the political offer of leading formations such as the Northern League (Barcella 2018), there has always been an everyday internal and external racism practiced by common people (Naletto 2009).

Thus, given the different contexts, and in spite of the superficial resemblances, the meaning of civility discourses in different parts of the country is rather nuanced. In the case of Messina, though, the local element posed at the basis of the current declinations of this theme is, probably, the 1908 catastrophe. For the purpose of a contribution to the study of undesired events, it is therefore possible to argue that ruinous occurrences can rewrite both the material positions of the places and their self-representations. The socially constructed nature of calamities lies, besides their triggering causes, often imputable to technical choices and other human reasons, on the cultural dynamics that they are able to set in motion. So that certain collective tragedies are, among other things, cultural occurrences that determine complex public and political effects. Disasters, for example, make apparent what is latent. As in the case mentioned earlier, the positivistic and intimately racist cultural climate and discourses that covered Italy, and seemed not to be applicable to urban centers like Messina, all of a sudden become the only way through which a place gets to be known and told in the public sphere—up to the point that these recently acquired labels shaped the identity of a place and contributed to create new social classes and hierarchies.

2.3 Blaming the Victims

Yet, at a general level and for the sake of an analysis of the social responses to disasters, it is also interesting to notice how the discourses of the national press in the aftermath of the 1908 earthquake reveal the persistence of those mechanisms of blaming that characterize highly "mediatized" events even in today's world. In this respect, Van Dijk (1991, p. 193) has observed that, usually, public discourse blames the victim for having acted in ways that justify prejudice or unequal treatment. Davis and French (2008, p. 249 passim) in their analyses of the discourses surrounding Hurricane Katrina have observed that media reports imbued victims and survivors with particular racial and class characteristics—namely, in the case at hand, being black and poor. Then, quite predictably, media linked victims and survivors to criminal activities. In third place, victims and survivors were portrayed as irresponsible. Yet, contrasts and divisions were utilized to create "good" and "bad" survivors. The substance of this analysis as well as these insights on the deployment of narrative devices aimed at imbuing the victims of Hurricane Katrina with particular stereotypes and assumptions related to their "identity", and at contrasting and dividing this "collective character", also apply in the case of Messina of a century earlier. If, as Caminiti (2010) and Centorrino (2010) have shown, the stereotypes by means of which the survivors of the earthquake were portrayed were consistent with the Italian ideology of the time, the way the media operated was quite modern. This is certainly true, among other things, even from a technical point of view, as the Messina earthquake is one of the first examples of organization of integrated and internationalized media networks able to cover an event of this sort (Castronovo and Tranfaglia 1979)—an occurrence that had clear implications for the way professionals active in the field interpreted reality and exposed themselves to unwitting biases (Centorrino 2010, p. 229). But, above all, media "modernity"—something, indeed, that bodes for the static nature of the bourgeoisie press in the Italian context, as is shown later in the book—concerns the narrative aspects and the way such narrations were aimed not simply at informing, but at expressing an ideology and confirming certain beliefs on the South and its people.

2.4 Public Memory and Latent Forces in Local Society

In the light of what we said before, it is possible to argue that tragic collective events augment the social reality by increasing the visibility of social structures and hidden tendencies. Public memory—that is, the circulation of recollections that entail what is remembered by the communities and how it is framed (Houdek and Phillips 2017)—shows these processes. For example, the Chicago Great Fire of 1871 took place within the framework of an ascending economy, and it can be seen through the lens of the myth of creative destruction. For the collective memory, from the Great Fire, the city resurrected stronger than before (Sawislak 1995). The same is true for San Francisco less than a decade after the 1906 earthquake. In 1915, the magnificent Panama Pacific International Exposition held in the Californian city portrayed the disaster as a morality play in which the courage of the inhabitants led to the "miracle of restoration" (Dyl 2017, p. 263). The common element between these two experiences is that, prior to the different disasters that hit them, both cities were in an expansion phase. On the contrary, in 1908 Messina presented a strong but declining economy (Battaglia 2010, p. 63 passim). As a matter of fact, since the 1880s, the agricultural sector, in particular wine and citrus fruit, had dipped consistently. The same was true for the textile industry. The silk business, in particular, was in crisis due to epidemics that had devastated the local silkworm rearing sector, and to the increasing competition of the Asian markets. Moreover, if for long time the port of Messina had been a hub for the goods coming from Palermo and Catania, at the beginning of the twentieth century new ports had emerged in the island, which were much closer to the regional sites of production of exportable goods. The building of new railway lines and other infrastructures benefited other provinces but not Messina, which saw its logistic relevance decreasing considerably. Yet, if for centuries Messina had been a free port, since the early 1880s that privilege had been abolished. Finally, at the turn of the century, citrus fruits and derived products were almost the only goods that left the port of the city—something that suggested the limitedness of the local economy and its upcoming collapse (Battaglia 2010; Chiara 2010).

Yet, if it would not be correct to state that the reconstruction began in 1924—that is, sixteen years after the earthquake—it is quite accurate to maintain that it had an acceleration starting from that year. The beginning of the Great War, the embezzlement or the diversion of the funds destined for the restoration, the inefficiency of the bureaucratic machine, and the particular political role of the devastated city had halted the reconstruction of the Sicilian center for almost two decades. And while the main reason for the decadence of the city had clearly been the earthquake, other endogenous and exogenous factors lay behind this process. These latter were represented by the partial disappearance of the core component of the local bourgeoisie, the nepotism of the local and national cliques, and the speculative processes fed by the politicians, which left those who lacked of economic and social capital without any type of guarantee (De Pasquale 2006). Due to prevalence of these local, endogenous factors, it is possible to claim that the earthquake had only accelerated the deterioration of the city, and that the decay was, so to say, around the corner anyway. Following Gotham & Grenberg (2014, p. 16), in fact, it is possible to maintain that only rarely disastrous events are the fundamental cause of urban crises. However, as a consequence of these different combinations of factors at work, at a cultural level, the Chicago Fire and the San Francisco earthquake generated internal narrations about rebirth, power, and determination; the Messina earthquake, instead, still evokes the ideas of end, misery, and nostalgia.

2.5 Public Responses to Disaster

However, beyond public memory, there are other tell-tale signs that make it possible to infer the levels of marginality, or the centrality of places hit by calamities—for instance, the quality of State response in post-disaster phases. While the acute phases of an undesired event is—especially in highly mediated environments as the modern one—a performative time whose outcomes depend on many elements, quality and timing of post-reconstruction are instead the elements that reflect the nature of power relationships active in the field. In the case of Messina, it is possible to claim that the quality of interventions in the acute phase was generally very bad and that it has been even worse in the post-disaster phase. Let's leave aside that the Italian rescue teams reached Messina only after several days (after the Russian and the British navies). And that in at least one case, the Italian rescuers had forgotten tents and blankets for the survivors. Or, that the name of the coordinator of the operations, General Francesco Mazza, has changed forever the Italian language by becoming synonymous with stupid. And even that the Italian troops appeared more concerned with rescuing bank vaults than people (Boatti 2004). What is relevant for our reasoning, instead, is that post-disasters' outcomes may depend on the position of a location of disaster in the national geography of power. Not only in the obvious sense that the political proximity between local and national governments can help mobilize resources for reconstruction, as the recent case of the 2012 earthquake in Emilia Romagna has shown (in a time during which the regional and central governments were both Democratic), nor in the sense that disasters that take place in economically strategic locations are more likely to see effective responses (Nimis 2009; Pitzalis 2016). The case of Messina, rather, shows how interventions in calamitous scenarios can represent a way to establish political orders in places that are not aligned with the ruling government in a given moment. On the eve of the earthquake, the economic choices of the Giolitti government had harmed Messina. Subsequently, the representatives of the city and the province in the national parliament had joined the opposition. And even at a local level, changes in the composition of the city council showed that the city was moving against the national government, raising, among other things, the interest of the national press (Messina was not a "city of peasants" yet, and certain changes were indeed likely to cause national political crises) (Cicala 2010, p. 426).

Three months after the earthquake, in March 1909, there would have been the national elections. Over the course of the time that preceded the vote, the then-Prime Minister Giolitti undertook many legislative initiatives in favor of the city. Extraordinary funds were allocated by way of derogation from the preexisting regulations; tax exemptions for the areas hit by the quake were introduced; the national economic development agency, *Cassa depositi e prestiti*, was allowed to grant depreciable loans, repayable in fifty years, to the owners of damaged houses. The enactment of such extraordinary legislative measures, together with many others, led up to Giolitti's nomination in the constituency of Messina, where, in fact, he emerged triumphant in spite of the many failures and delays that had characterized remediation until that moment. These acts were part, in other words, of a strategy aimed at taking control of the city again and promoting the image of the government (Noto 2008, p. 208).

One year after the elections, in 1910, Giolitti's government issued the law n. 466 through which it created the *Unione Messinese dei Danneggiati dal Terremoto*, or the union of those who had been harmed by the Messina earthquake. The *Unione*, whose responsibilities grew over the course of time, was in charge of gathering all the owners of damaged houses and coordinating the restoration, or reconstruction, of the buildings to be erected on the lots under control of the agency itself. Members of the *Unione* would have qualified for loans amortizable in thirty years, for a value corresponding to 75% of the value of the old houses. In practice, this measure denied the right to housing to those who did not own a house before the tragedy, or not possessed the means required to repay their loans. As is discussed in the following chapters, those who did not qualify for a loan found themselves forced to live in those barracks which will become part of the city landscape for a century to come, giving rise to that "social question" that is at the very center of the present book.[2] Yet, being the loans granted in proportion to the value of the harmed house, this system favored big owners. Finally, the loans were transferrable, and this paved the way to big speculations—so jeopardizing both the lower and the middle classes (Noto 2008, pp. 228–230). From this moment on, the strains and the conflicts that had previously characterized the relationship between the local elites and Giolitti were temporarily eased by this new convergence of interests: the same convergence that opened up the way to that "capitalist use of the earthquake" (Barone 1982) which is at the very basis of the current urban social stratification. However, such capitalist use of the earthquake was not limited to the local level. As a matter of fact, Giolitti's strategy opened the door also to the northern cooperatives and the big enterprises that operated, and often still operate, in the real estate and the infrastructure sector. Not by chance, very soon, the *Unione* would have become a national organ responsible for the assignment of funds in case of disaster, and Giolitti himself would have been elected again in the same district in the elections scheduled for 1913. In this respect, not even the scandals could halt the prime minister. For example, a scandal that exploded in 1912 showed that the *Prefetto* of Messina—that is, the representative of the State in each administrative province—and the Minister of Public Works, had engaged in a conflict over the selection of the companies which would have been awarded public procurement contracts for carrying out certain public works. The *Prefetto* appeared to support the local enterprises, and the Minister the northern cooperatives. The inquiry demonstrated that at the end of the day, the two authorities had found a way to settle the dispute and award the contracts to both parties (Noto 2008, p. 233).

In sum, beyond his political goals, Giolitti's uses of the disaster set in motion shared predatory mechanisms, aimed at the extraction of value from both the land and the ruinous events, which are very modern and contemporary. Klein's notion of disaster capitalism—meant as the "orchestrated raids on the public sphere in the wake of catastrophic event, combined with the treatment of disasters as exciting market opportunities" (Klein 2007, p. 6)—applies very well to the case at hand. In the years that followed the earthquake, Messina presented many such opportunities; moreover, the liberal ideological framework as well as the legal and administrative tools utilized to trigger the exploitative process, resembled those identified by this author in her analysis. But if Klein traces the origin of a certain model of governance in events and policies which made their first appearance in the 1970s, our hypothesis is that the beginning of disaster capitalism has older roots.[3] And that, more than to neoliberalism, speculations, abuses, and violent value extraction pertain to capitalism and, in particular, to those states that present a developed financial system, an extended bourgeoisie, a large territory, a complex and decentralized bureaucracy, and a legal system that allows "exceptions".

2.6 The "Productive" Role of Disasters

However, a reflection on the "positioning" of a city within a national political scenario in the aftermath of a calamity implies also a discussion of the positions internal to the city—that is, on the complex scalar levels of dependency present in a place. In Tsing's terms, scalability "is the ability of a project to change scales smoothly without any change in project frames" (Tsing 2015, p. 38). Or, "so above, so below", as it is suggested by one of the most important credos of that hermetic philosophy whose mention is pertinent in this context both for its connections to the concept of scale and for being at the base of the ideology of an Italian political class that, above and below, in Messina and in the rest of Italy, was at that time largely masonic and, thus, gathered around narrow circles characterized by affiliation, reciprocity, and interests (Fedele 2009; Conti 2014). Dependency, yet, is a central concept that implies in our case both the intertwinement of the interests held by the elites (politicians, professionals, and entrepreneurs of local and national level) and the asymmetry between actors.

In our case, the term dependency points both at the forced connection that links the semi-periphery to the political center (the national State) and at the interpersonal reliance by the lower classes on the upper ones, by means of brokers and cliques (Chubb 1982). A semi-peripheral dependency is the condition experienced by Messina after the seismic waves had shattered the city, together with its infrastructure, investors, and wealth. If it is not credible that this economic and social capital was entirely gone and that the city had no specific weight after the event, certainly its importance was at this point considerably downsized.

For the purposes of our analysis of the functions of disasters, these concepts are important because they point to the "productive" role of calamities. These events and, above all, the management processes that follow, do not produce only public memories or identities—that is, real but at the same time abstract notions concerning the collective Self of a place. Rather, they generate social relations and practices—like, among others, Barrios (2017) has shown in a portion of his study devoted to the relation between Hurricane Mitch and the rise of the *mara*, the gangs, in Honduras. In our example, Messina, like most of the South of Italy, is famous for being home to nepotistic and clientelistic relations that are distinguished from but are also connected to that quintessential local mode of regulation which is called mafia (Walston 1988). This is very evident in the Messina case. Certainly, the city was considered *babba*, non-mafiosi, as we said earlier, but only on the surface. Indeed, over the years, a number of investigations showed how, for decades, Messina has been firmly part of the most important criminal networks that operated on the island and elsewhere (Comitato messinese per la pace e il disarmo 1998; La Rosa 2013). And those studies have also shown how the mafia active in the city crossed the social classes by involving and connecting members of the bourgeoisie and the proletariat.

The complex and differentiated links between clientelism, political power, private entrepreneurship, and the mafia developed in the interconnection between local, national, and international networks. Moreover, as we have already mentioned, they have often been simplified in the positivistic and stereotyped discourse on the southern character (idle, indifferent, and violent), in which the mafia (as well as brigandage) was seen as a natural complement of this feral nature (Dickie 1992; Riall 1998).

Given this layered context of meanings, attributions, and societal positions, the Messina disaster was, as we said, "productive" for it has showed how these cliques as well as the "intersectional" memberships (bourgeoisie and mafioso, mafioso and man of respect well accepted in the circles of the upper classes) and the other new hierarchies active in the city had been generated by the State and, in particular, by the tactical choices of the central government under Giolitti's control. This is not to say that before the earthquake, these tendencies were not present in city life. On the contrary, well before the earthquake, a local witness of the time, the engineer Riccardo Hopkins, was usually very vehement against the cliques that ruled the city (Hopkins 1882). Likewise, historians are keen to concede that the influence of this composite group of politicians and businessmen, not foreign to violence and intimidations, was indeed quite strong (Caffo 2008). However, it is indeed noticeable that in the moment in which the eyes of the central authorities were directed toward the city as never before in the brief history of the unitary State, this very same State did not try to eradicate those practices that had qualified a social space as immoral and dysfunctional. On the contrary, it employed these practices, forces, and dynamics to stabilize the environment, and produce, locally, that is "below", that very social order that, "above", at a central level, had allowed a national "block of interests" composed of patrons and clients to become a State. Conversely, this new order broadened that mass of beggars and "needy" classes that, in turn, joined the army of clients and, possibly, of members of organized crime.

It was indeed a complex and scalar architecture of power whose levels of premeditation, perhaps, should not be exaggerated. As a matter of fact, the production of a vast underclass presented some risks, given the proximity of the adoption of universal suffrage in 1912 and the national elections that took place the following year—especially in the presence of a growing socialist movement and, above all, of a Catholic party, which, in fact, would go on to triumph in 1913 both locally and nationally. There was, hence, a sort of foolishness in this way of operating that does not bode for any rationalist interpretation, but reflects the substantial negligence of the elites of power, together with their intimate idea that only business, and the upper classes, mattered. However, whether initially this process of creation of a mass of shanty-dwellers, which in 1922 consisted of 70,000 people (Noto 2008, p. 239), was intentional or not, over the decades, it resulted in a precise mode of domination: a clientelistic mode that has been organized by the central State and nourished by national and local players for a century to come—as shown in the coming chapters.

2.7 Bureaucratic Classifications and Social Responses

Another implication for the study of disasters, therefore, is that these events should entail a study of the bureaucracies as zones of ambiguity and uncertainty, within which the actors exercise a discretionary power, with uncertain results (Crozier 1964; Crozier and Friedberg 1980). Either impersonal organizations that operate in the general interest, or structures that can be worked out, bureaucracies can add to the deadly effects of a disaster and reflect active differences in access to services, licenses, and subsidies. On the traces of Douglas (1986), in her study of a vast landslide that hit a number of Sicilian villages in 2009, Falconieri (2017, p. 137 passim) has shown how the classificatory activity of the bureaucracies in the case of disaster—an activity that consists, among other things, in assessing the value of damages or conferring provisional shelters—is far from being a neutral operation. Often, this operation reflects values and schemes that might not be shared by the actors—that is, the authorities and the citizens. It can revitalize old fractures and lines of separation that are at the very basis of the processes of national formation (Holmes 2000; Herzfeld 2005): public affects and sentiments, in other words, which are directed against the State and the national rhetoric. The latter is an occurrence that emerges with much clarity in the abovementioned ethnographic works of Falconieri (2017) and Hochschild (2016). For different reasons, ultimately connected to institutional classifications and discourses, in the cases studied by these scholars, the communities hit by a disaster have developed self-exclusionary and isolationist sentiments, which have the "government" or the "country" as a target. A feeling of estrangement which is also due to the fact that bureaucracies can redraw the geography of place and affects, as, among many others, the studies of Barrios (2017) on the social effects of Hurricane Katrina in New Orleans, and the research collective Emidio di Treviri (2018) on the 2016 earthquakes in central Italy, have demonstrated.

Moreover, like the case of Messina in 1908 suggests, bureaucracies can work in the general interest or in the interest of particular groups of power. Especially under regimes of exception and emergency, the uncommon capability of certain cliques and individuals to have access to public offices is often among the reasons behind the rise of new subjects (entrepreneurs, firms, speculators, etc.), the marginalization of others, and the new configuration of space. Processes of dispossession, though, do not take place without open or subtle forms of opposition and adaptation. Benadusi (2012), for instance, has shown how a poor Sri Lankan community that had been excluded by the international donations, in the aftermath of the 2004 tsunami, put into being a collective social drama aimed at obtaining part of the funds. Londero (2008) has described how a small community in Friuli, hit in 1976 by one of the most terrible earthquakes of contemporary Italy, had gathered in a tent camp and, from there, had imposed with its struggles the best reconstruction model even witnessed in Italy. A similar story is narrated by Pitzalis (2016), whose ethnographic study focuses on a leftist community in Emilia Romagna that had eschewed the governmental camps of the Civil Protection Department in order to create an autonomous space, from which they conducted a struggle against the policies of the central and regional governments. Carnelli (2015), also, showed how a group of activists of L'Aquila, the Italian city destroyed by a disastrous earthquake in 2009, reinvented religious traditions and organized a procession in order to violate the "red zone", that is, the vast urban area interdicted to the citizens.

But while these are examples of open and collective opposition to reconstruction policies, practices of resistance and adjustments are often hidden and individualistic. This latter is the case with the seventy-seven people that, in the Sicilian villages of Scaletta and Itala studied by Falconieri (2017, p. 131 passim), were sued for having violated a norm that impeded them from using, or even opening, their partially damaged houses after they had been forcibly evicted and, above all, at least in some cases, forced to receive subsidies to housing. None of these people, according to the intimate account of Falconieri, had taken advantage of the measures issued by the authorities or had acted against the spirit of the law, in spite of their personal disapproval of the decisions imposed by the authorities. All of them, rather, had used their old houses only for a few hours every day, either for storage or as a place within which to exercise some kind of spiritual reconnection with their lost lives.

Still different is the case of the young farmers studied by Breglia et al. (2018) in Central Italy. Unlike the older generation of peasants and herders that was tremendously affected by the 2016 earthquake, this new cohort of better-educated farmers, able to read and interpret the hundreds of by-laws and calls for support grants issued by a number of agencies in the aftermath of the calamity, were capable of transforming this profound event into an opportunity and so increasing considerably their profits. Something, incidentally, that allows us to see in a critical perspective the mainstream discourse on resilience—a concept that, often, neglects the importance of asymmetries within studied contexts.

As noted by Hull (2012, p. 1), societies are surrounded and controlled by a flow of bureaucratic documents, whose materiality has vast consequences and are constitutive of individual and collective stories. An ethnography of the material production of public administration, then, is a way to see the world through the lens of the State. While an ethnography of the people in this same bureaucratic arena is, as always, a way to observe the world through the mirrored perspective of those who are caught, and also "produced", by the particular sight of the administration (Scott 1998). An analysis of the bureaucracy, therefore, is in the case of modern calamities a necessary way to explore the complex interplay between the strategies of the State and the responsive tactics of the common people to oppose regulation. The comparison among bureaucratic styles, finally, is the tool to understand current trends and images of desirable modernity with regard to urbanism, public space, political economy, and city management (Gotham and Greenberg 2014).

After all, as argued by Powers (2006, p. 32), history has shown that the recovery from disasters is largely a reflection of the values and commitments of those charged with restoring an area. Values and commitments, we add, which are shared by segments of the economic, political, and intellectual elites in a given time, and that, at the end of a struggle of ideas and interests, get to distinguish different epochs or places, producing different outputs in terms of social inclusion and regeneration. Nevertheless, in spite of different results, and regardless of their degree of inclusion, these styles and values share the tendency to identify enemies and generate discourses aimed at producing deviants and scapegoats. From looters to bad commanders in order, the diffusion of myths, religious beliefs, symbols, and plot theories shows that calamities are highly "culturalized" events, which not even the highly rational logic of the State can escape—what, in Rock's terms (2013, p. 42), can be defined as a supplementary set of "pseudo-disasters" that take place within real calamitous scenario and point at the very obvious fact that undesired events are, ultimately, a matter of order.

From this observation, finally, derives the cynical note that disasters are the perfect ethnomethodological situation. That is, in Garfinkel's words (2011, pp. 54–55), the time and the place in which familiar environments lose their common background and become specifically senseless. In this situation, bewilderment, uncertainty, internal conflict, psycho-isolation, and acute and nameless anxiety are the feelings that prevail. Within this context, then, the only question that matters to people is how to re-create a world. Disasters, therefore, like the conditions created in the famous experiments by Garfinkel, are those rare, natural situations that allow people and observers to see latent social forces emerge spontaneously, together with people's attempts to reconstitute their life worlds and assign meanings to things. The most essential and (pre-)political of the situations one can imagine.

Notes

1. Zampieri (2018) is the only author who has underlined the importance of this distinctive category for the local culture.
2. A similar point, on how economic injustice in the USA has been exacerbated through housing policy, is made by Rothstein (2017).
3. Probably, such origins might be dated back to the eighteenth century. At least if, to remain in the field of disasters research, one follows, among many authors and studies on modern disasters, Placanica's research on the 1783 Messina earthquake (Placanica 1985, pp. 162–164) or Pereira's economic analysis of the 1755 Lisbon's earthquake (Pereira 2009, p. 496). Albeit these two reconstruction processes were managed incomparably better than the seismic crisis of 1908, one will find the seeds of those tendencies whose origins are traced by Klein back to the capitalist restructuring processes of the early 1970s. The elements of institutional violence (the use of the army, in particular); the implementation of states of exception; the raise of new individuals and groups, or the use of disasters as a way to impose radical changes in the State organization and the law, can be found also in those experiences. Certainly, this hypothesis should be deepened. And one of the problems lies in the fact that many "minor" disasters have been discussed mostly in the national languages and are thus relatively difficult to know. But, certainly, there would be much to learn from those events that took place at the "outskirts of the empire", and they remain virtually unknown to most observers.

References

Barcella, P. (2018). Percorsi leghisti. Dall'antimeridionalismo alla xenofobia. *Meridiana, 91*, 95–119.

Barone, G. (1982). Sull'uso capitalistico del terremoto. Blocco urbano e ricostruzione edilizia a Messina durante il fascismo. *Storia Urbana, 10*, 47–104.

Barrios, R. (2017). *Governing Affect. Neoliberalism and Disaster Reconstruction*. Lincoln: University of Nebraska Press.

Battaglia, R. (2003). *L'ultimo "splendore". Messina tra rilancio e decadenza*. Soveria Mannelli: Rubbetino.

Battaglia, R. (2010). Messina tra crisi e tentativi di rilancio dal "lungo Ottocento" al terremoto. In A. Baglio & S. Bottari (Eds.), *Messina dalla viglia del Terremoto del 1908 all'avvio della ricostruzione* (pp. 57–70). Messina: Istituto di Studi Storici Gaetano Salvemini.

Benadusi, M. (2012). The Politics of Catastrophe. Coping with 'Humanitarism' in Post-Tsunami Sri Lanka. In F. Attina (Ed.), *The Politics and Policies of Relief, Aid and Reconstruction. Contrasting Approaches to Disasters and Emergencies* (pp. 151–172). New York: Palgrave Macmillan.

Boatti, G. (2004). *La terra trema*. Milano: Mondadori.

Breglia, A., Chiloiro, A., Inserra, G., et al. (2018). Le faglie della terra. Uno studio sul mondo rurale nel post-sisma dell'Appennino centrale. In E. di Trevidi (Ed.), *Sul fronte del sisma. Un'inchiesta militante sul post-terremoto dell'Appennino centrale (2016–2017)* (pp. 204–229). Rome: DeriveApprodi.

Caffo, A. (2008). Politica e Massoneria a Messina. In G. Bolignani (Ed.), *Francesco Durante. Politico e massone* (pp. 25–73). Messina: Istituto Novecento.

Caminiti, L. (2010). "La mela avvelenata". Efficientismo e pregiudizi negli aiuti alle popolazioni terremotate. In A. Baglio & S. Bottari (Eds.), *Messina dalla viglia del Terremoto del 1908 all'avvio della ricostruzione* (pp. 175–794). Messina: Istituto di Studi Storici Gaetano Salvemini.

Carnelli, F. (2015). La festa di San Giovanni a Paganica. Riti e Santi fra le macerie del post-sisma aquilano. In P. Saitta (Ed.), *Fukushima, Concordia e altre macerie. Vita quotidiana, resistenza e gestione del disastro* (pp. 135–148). Firenze: Editpress.

Castronovo, V., & Tranfaglia, N. (1979). *La stampa italiana nell'età liberale*. Bari: Laterza.

Centorrino, M. (2010). Il terremoto e le dinamiche della rappresentazione giornalististica. In A. Baglio & S. Bottari (Eds.), *Messina dalla viglia del Terremoto del 1908 all'avvio della ricostruzione* (pp. 223–245). Messina: Istituto di Studi Storici Gaetano Salvemini.

Chiara, L. (2010). I patrimoni e le strategie d'intrapresa economica della borghesia messinese (1880–1908). In A. Baglio & S. Bottari (Eds.), *Messina dalla viglia del Terremoto del 1908 all'avvio della ricostruzione* (pp. 71–81). Messina: Istituto di Studi Storici Gaetano Salvemini.

Chubb, J. (1982). *Patronage, Power, and Poverty in Southern Italy: A Tale of Two Cities*. Cambridge: Cambridge University Press.

Cicala, A. (2010). Blocchi fulciani e clerico-moderati. Lotte politiche a Messina dopo il terremoto del 1908. In A. Baglio & S. Bottari (Eds.), *Messina dalla vigilia del Terremoto del 1908 all'avvio della ricostruzione* (pp. 425–442). Messina: Istituto di Studi Storici Gaetano Salvemini.

Comitato per la pace e il disarmo (Ed.). (1998). *Le mani sull'Università. Borghesi, mafiosi e massoni nell'ateneo messinese*. Messina: Armando Siciliano Editore.

Conti, F. (Ed.). (2014). *La massoneria italiana da Giolitti a Mussolini*. Rome: Viella.

Crozier, M. (1964). *The Bureaucratic Phenomenon*. Chicago: University of Chicago Press.

Crozier, M., & Friedberg, E. (1980). *Actors and Systems. The Politics of Collective Action*. Chicago: The University of Chicago Press.

Davis, M. J., & French, T. N. (2008). Blaming Victims and Survivors: An Analysis of Post-Katrina Print News Coverage. *Southern Communication Journal, 73*(3), 243–257.

De Martino, E. (2002). *La fine del mondo*. Torino: Einaudi.

De Pasquale, D. (2006). *Le mani su Messina. Prima e dopo il terremoto del 1908*. Messina: Armando Siciliano Editore.

Di Treviri, E. (Ed.). (2018). *Sul fronte del sisma. Un'inchiesta militante sul post-terremoto dell'Appennino centrale (2016–2017)*. Rome: DeriveApprodi.

Dickie, J. (1992). A Word at War: The Italian Army and Brigandage 1860–1870. *History Workshop Journal, 33*(1), 1–24.

Douglas, M. (1986). *How Institutions Think*. Syracuse, NY: Syracuse University Press.

Dyl, J. L. (2017). *Seismic City. An Environmental History of San Francisco's 1906 Earthquake*. Seattle: University of Washington Press.

Falconieri, I. (2017). *Smottamenti. Disastri, politiche pubbliche e cambiamento sociale in un comune siciliano*. Rome: Cisu.

Fedele, S. (2009). Le logge massoniche di Messina alla vigilia del terremoto. In G. Campione (Ed.), *La furia di Poseidon. Messina 1908 e dintorni* (pp. 115–124). Palermo: Silvana Editoriale.

Fofi, G. (1964). *L'immigrazione meridionale a Torino*. Milano: Feltrinelli.

Foot, J. (2001). *Milan since the Miracle. City, Culture and Identity*. Oxford and New York: Berg.

Garfinkel, H. (2011). *Studies in Ethnomethodology*. Cambridge: Polity.

Gibson, M. (2002). *Born to Crime: Cesare Lombroso and the Origins of Biological Criminology*. Westport, CT: Praeger.

Gotham, K. F., & Greenberg, M. (2014). *Crisis and Redevelopment in New York and New Orleans*. New York: Oxford University Press.

Herzfeld, M. (2005). *Cultural Intimacy*. New York: Routledge.

Hochschild, A. R. (2016). *Strangers in Their Own Land. Anger and Mourning on the American Right*. New York: The New Press.

Hofstadter, R. (1955). *The Age of Reform*. New York: Vintage Books.
Holmes, D. (2000). *Integral Europe. Fast-Capitalism, Multiculturalism, Neofascism*. Princeton: Princeton University Press.
Hopkins, R. (1882). *La vera piaga di Messina*. Messina: Tipografia del Faro.
Houdek, M., & Phillips, K. (2017). Public Memory. *Oxford Research Encyclopedia of Communication*. Retrieved from http://communication.oxfordre.com/view/10.1093/acrefore/9780190228613.001.0001/acrefore-9780190228613-e-181.
Hull, M. S. (2012). *Government of Paper. The Materiality of Bureaucracy in Urban Pakistan*. Berkeley and Los Angeles: University of California Press.
Huysseune, M. (2006). *Modernity and Secession. The Social Sciences and the Political Discourse of the Lega Nord in Italy*. New York: Berghahn.
Italiano, F., Bonfanti, P., Ciaccio, C., Gesù, M., Romano, C., Schiavone, S., & Piazza, G. (2008). Degassamento di radon e modifica del DNA della popolazione siciliana: possibili relazioni con il terremoto del 1908. *Ingv Conference Materials*. Retrieved from https://www.earth-prints.org/handle/2122/5028.
Klein, N. (2007). *The Shock Doctrine. The Rise of Disaster Capitalism*. New York: Henry Holt and Company.
La Rosa, M. (2013). *Il fenomeno Mafioso. Il caso Messina*. Messina: Armando Siciliano Editore.
Londero, I. (2008). *Pa sopravivence. No pa l'anarchie. Forme di autogestione nel Friuli terremotato. L'esperienza della tendopoli di Godo (Gemona del Friuli)*. Udine: Forum.
Moe, N. (2002). *The View from Vesuvius. Italian Culture and the Southern Question*. Berkeley and Los Angeles: University of California Press.
Montaldi, D., & Alasia, F. (1960). *Milano, Corea. Inchiesta sugli immigrati*. Milano: Feltrinelli.
Naletto, G. (2009). *Rapporto sul razzismo in Italia*. Rome: Manifestolibri.
Nimis, G. P. (2009). *Terre mobili. Dal Belice al Friuli, dall'Umbria all'Abbruzzo*. Rome: Donzelli.
Noto, A. G. (2008). *Messina 1908. I disastri e la percezione del terrore nell'evento terremoto*. Soveria Mannelli: Rubbettino.
Oliver-Smith, A. (2002). Theorizing Disasters: Nature, Power, and Culture. In S. M. Hoffman & A. Oliver-Smith (Eds.), *Catastrophe and Culture. The Anthropology of Disaster* (pp. 23–47). Santa Fe: School of American Research Press.
Pereira, A. S. (2009). The Opportunity of a Disaster: The Economic Impact of 1755 Lisbon Earthquake. *The Journal of Economic History, 69*(2), 466–499.
Petraccone, C. (2000). *Le due civiltà. Settentrionali e meridionali nella storia d'Italia*. Rome and Bari: Laterza.
Pitzalis, S. (2016). *Politiche del disastro. Poteri e contropoteri nel terremoto italiano*. Verona: Ombre Corte.
Placanica, A. (1985). *Il filosofo e la catastrofe. Un terremoto del Settecento*. Torino: Einaudi.

Powers, M. P. (2006). A Matter of Choice: Historical Lessons for Disaster Recovery. In C. Hartman & G. D. Squires (Eds.), *There Is No Such Thing as a Natural Disaster* (pp. 13–34). New York: Routledge.

Pruiti, R. (2010). *La città negata. Viaggio nelle baraccopoli di Messina a un secolo dal terremoto*. Rome: L'Espresso.

Rabinow, P. (1995). *French Modern. Norms and Forms of Social Environment*. Chicago: The University of Chicago Press.

Redazione. (2016). Il terremoto del 1908 ha cambiato l'identità dei messinesi?. *Scirokko.it*. Retrieved from http://scirokko.it/il-terremoto-del-1908-ha-cambiato-lidentita-dei-messinesi/.

Riall, L. (1998). *Sicily and the Unification of Italy: Liberal Policy and Local Power*. Oxford: Clarendon Press.

Rock, P. (2013). *Deviant Behavior*. London: Routledge.

Rothstein, R. (2017). *The Color of Law. A Forgotten History of Our Government Segregated America*. New York and London: Liveright Publishing.

Roy, A. (2004). The Gentleman's City: Urban Informality in the Calcutta of New Communism. In A. Roy & N. Alsayyad (Eds.), *Urban Informality. Transnational Perspectives from the Middle East, Latin America, and South Asia* (pp. 147–170). Lanham, MD: Lexington Books.

Roy, A., & Alsayyad, N. (Eds.). (2004). *Urban Informality. Transnational Perspectives from the Middle East, Latin America, and South Asia*. Lanham, MD: Lexington Books.

Saugeres, L. (2002). Of Tractors and Men: Masculinity, Technology and Power in a French Farming Community. *Sociologia Ruralis, 42*(2), 143–159.

Sawislak, K. (1995). *Smoldering City. Chicagoans and the Great Fire. 1871–1874*. Chicago: The University of Chicago Press.

Schneider, J. (Ed.). (1998). *Italy's 'Southern Question'. Orientalism in One Country*. New York: Berg.

Scott, J. C. (1998). *Seeing Like a State. How Certain Schemes to Improve the Human Condition Have Failed*. Yale: Yale University Press.

Tsing, A. L. (2015). *The Mushroom at the End of the World. On the Possibility of Life in Capitalist Ruins*. Princeton: Princeton University Press.

Van Dijk, T. A. (1991). *Racism and the Press. Critical Studies in Racism and Migration*. London: Routledge.

Walston, J. (1988). *The Mafia and Clientelism: Roads to Rome in Post-War Calabria*. London: Routledge.

Zampieri, P. (2018). *Esplorazioni urbane. Urban art, patrimoni culturali e beni comuni*. Bologna: Il Mulino.

CHAPTER 3

History Seen Through the Slums: The Southern Question and the Current Crisis

3.1 Messina, No South and No North

The previous chapter began with a rather provocative question: was the Messina earthquake a disaster or a catastrophe, if we consider the latter an irreversible event? (Noto 2008).

In the collective memory of the residents, the earthquake was an irreversible event, perhaps a permanent and irremediable rift. Whether or not this corresponds to reality or to the imagination is unimportant. What really matters is that this constant reiteration, practice, and reproduction generate real effects.

Photographs showing the city reduced to a pile of rubble seem to show the "end of the world" (De Martino 2002; see Chap. 2), or perhaps the end of "a world", and engender the desperate and romantic stories of a mythical lost city replaced by slums with no past/memory,[1] that are in constant need of restoration but cannot be restored. As is seen in more depth in Chap. 5, it is within this semantic parameter that the space of "cultural intimacy" is constructed (Herzfeld 2005).[2] This lies at the center of collective and individual identity strategies that come into play both inside and outside the city, in a form of reciprocal interaction between the *insiders* (citizens who belong to different social and spatial circles within the city) and the *insiders* and *outsiders*, when "the city" becomes part of the national political arena.

D. Farinella, P. Saitta, *The Endless Reconstruction and Modern Disasters*,
https://doi.org/10.1007/978-3-030-19361-4_3

Messina *is* the 1908 earthquake, in the sense that a collective, shared narrative of and about the city would not exist without this event, which, ambivalently, produces a sense of loss (*what was and no longer is*) that is "ontological"; in other words, there is a sense of displacement linked to "being in the world" (Giddens 1990, p. 92).[3]

At the same time, the numerous shacks, illegal self-built houses, and public housing scattered all over the city, and the ensuing urban confusion, are the noisy and abundant consequences that artificially fill the empty space left by the earthquake. They fill the space physically, but more especially, symbolically, as the perpetual existence of these buildings is the self-evident expression of the impossibility of ever returning to normal, of getting over the sense of loss.

An event such as an "earthquake" is not only a temporal watershed but also a cultural one; there is a *before* and *after* the earthquake. It is a material temporality in which *before* and *after* are two linear moments, both consecutive and continuous (so that the second necessarily contains the first), and that can be translated into an abstract, ahistorical, and recurring temporality: thus, *before* and *after* become monolithic blocks which exclude each other but are complementary, like two sides of the same coin. In the time *before* the earthquake, the city is seen as mythical, with its merchants and free port, as a city that played an important role in Mediterranean maritime trade (see Chap. 4); by deduction, the *after* is *what remains* beyond that mythical time. However, *what remains* is unquantifiable and does not follow a chronological, linear, and continuous pathway, but it repeats ad infinitum, spilling into the empty space that is left clamoring to be "filled". The resulting surface accumulation and disorderly juxtaposition (here, the slums are emblematic), which fill what is perceived as "empty" space to overflowing, simply confirms the sense of loss.

However, Messina is not simply in Italy, it is also in the South.[4] This geographical aspect is fundamental to the symbolic re-elaboration of its post-earthquake history. In this sense, Messina is part of southern history and bears the symbolic-ideological weight of the "Southern Question". In other words, it is a territory that has remained "behind", backward, and unable to keep pace with modernization, in a South that is the expression of material underdevelopment, as measured by the various socio-economic indicators (per capita gross domestic product [GDP], education levels, poverty rates, etc.), which underpin a social divide and huge cultural differences.

Within this framework, the persistence and tenacity of these "perpetual" shacks, which have existed since 1908 as part of an infinite "on paper only" reconstruction plan, are seen as just another expression of southern wastefulness, of the inability of part of the country to embrace modernity. The tension between a modernist policy based on rehousing and restoration projects (on which to build the good, clean and modern), as opposed to the "social cancer" of continuous "rehousing in other shacks" (the bad, dirty and backward) plunge the city into a distorted modernity: this is the Messina of the favelas, ***similar to*** those in Brazil or some African nations (placing Messina as part of the "Third world"), and the Messina of the ***shanty-dwellers who live like the Roma***.[5] In other words, it represents the scandalous contradictions of capitalism, or an Italy which ought—in official eyes—to be "civilized", part of a West that, whatever its past, still imagines itself through the positivistic myth of modernity, represented by progress and development.

How can this conflict be resolved? How can these contradictions be demonized? The most reassuring and placatory solution is the rather naive reality of finger pointing. The show of moralism implicit in seeking out the guilty party allows the accuser to rise to the role of "civilizer", thus underlining their distance from the problem, and allowing them to reduce the problem of the ***slums and their uncivilized and backward inhabitants*** to a "non-issue". As a result, the people reduced to living in those conditions become "different", they are "culturally" trapped, and, to use Pines' interesting metaphor (2012) regarding the representation and self-representation of the Neapolitans (whose issues also form part of the narrative concerning the South), are transformed into "garbage".

In this context, an interesting detail emerges from the "earthquake" in Messina; it is possible to date precisely when this cultural "change" took place to identify the chronological moment that created these differences. Following the fateful event, the seeds of the stereotypes of this area as characterized by resignation, dependence, welfarism, and the inability to take action were permanently established, committing it to a future of backwardness and underdevelopment in the national narrative.

Dickie (1999, pp. 143–144) had already observed: "the earthquake that devastated Messina and Reggio Calabria in December 1908 made the South the subject of what is probably Italy's first great cross-class movement of national solidarity. Yet at the same time, the disaster reinforced the image of a tragic South, an eternal reminder of the fragility of human progress".

Later, after Caminiti (2009) and Giarrizzo (2010), Gribaudi's (2010) analysis of the representations of "earthquakes" showed how the 1908 earthquake, together with those that followed in the rest of Italy, was part of the North/South dichotomy, so that:

> [after the earthquake] two completely different representations of the roles of Italians involved in disasters emerged: the North with its modern, generous, active, national rescue Committees; while the victims in the South were seen as passive, irrational, dirty troglodytes. Instead of being attributed to the inadequacy of public intervention and lack of coordination, inefficiency was attributed to the behavior of the victims: "as a result of the anti-social character of the southerners who were unable to live their lives according to modern values and standards". (Caminiti 2009, p. 104)

> The inhabitants' protests against the relocation of their villages and the restoration of their historic buildings were stamped by the authorities and the northern rescue committees as being the result of ignorance, of resistance to innovation by a population who were "naturally excitable and rather impressionable" (Ivi: 105), of people with an impetuous Mediterranean nature. The catastrophe "was, from the beginning, seen and experienced in a dualistic key: the North, who intervened in relief operations with prompt generosity while the South was uncivilized and backward, lazy and resigned (Giarrizzo 2010)". (Gribaudi 2010, pp. 85–86)

On the one hand, the earthquake in Messina is perceived, remembered, and narrated as an essentially natural, depersonalized fact, part of the powerful metaphor of man versus nature, in which the destructive forces of nature are seen to rebel against civilization, re-appropriating space and producing a sort of barbarization and "bestialization". In addition, by erasing the architectural-cultural (monuments, churches, and buildings) and social heritage (its trade and business families, its "active" middle class), its "History" and "Identity" are canceled out, so that the city of Messina returns to being a "paradise inhabited by devils".[6]

On the other hand, the nature/culture dichotomy becomes part of and strengthens the North/South dichotomy, placing Messina in what Herzfeld (2009, p. 131 e ss.) calls the symbolic geography of the Southern Question. The contradiction of a recovery and reconstruction process that is still ongoing today, while seemingly infinite, or even impossible, makes it impossible to see the city's history and the Southern Question as two separate entities. It also underlines all the ambiguities inherent in this concept, or rather, the cultural and essentializing trap, or "the stereotypical traits of backwardness, which initially serve to describe the conditions of the victim and increase feelings of pity, but can be reversed later on to explain the lack of reconstruction. The narrative is transformed. [...] The history of defeat focuses on the shortcomings, on the inability to act, on corruption" (Gribaudi 2010, p. 87).

Seen in the context of the "Southern Question", both Messina with "its earthquake" and "its shanty-dwellers" who have undergone a cultural transformation have become "illegal by profession",[7] thanks to the dual strategy of individual blame and the denial of institutional responsibility. They express the sort of "distorted" modernity stereotypically attributed to southern Italy, one that is seen as deficient and "unbalanced" (Barbagallo 1994) because it is trapped in a modernization process that is incomplete, backward, and "passive" (Cafagna 1988).

Development is "not autonomous", it is dependent and state-assisted (Trigilia 1992; Pezzino 1993), highlighting a "difficult modernization" "without growth" (see Davis 1998, p. 209 passim).

However, although Messina and its history are inscribed within the ideological framework of the Southern Question, this approach is limited, and it is necessary to look beyond the concept of South and North and to consider contemporary Messina as one of the many fragments of modernity.

The operation of cultural blame, which artificially transforms the 1908 earthquake victims into today's shanty-dwellers—the victims of a "slum" culture that has been internalized so that they are prisoners of social and urban deterioration—also serves to bolster the lack of institutional and political responsibility so typical of the way the Southern Question is perceived. As Herzfeld sums up, "the original sin of being a Southerner serves to absolve politicians and bureaucrats from any responsibility for the failure of the ideal transparent, rational, scientific, and non-corrupt state model" (2009, p. 143).

Furthermore, this simplified narration of the continued existence of these slums (they exist "because of" the earthquake, and have existed "since" the earthquake) places them in an extra-temporal time, or outside of history, making it difficult to carry out a critical analysis of the processes that have, over the course of the century, produced and reproduced the space in which the shacks and the rest of the urban settlement are located. We forget the "work" that took place over the course of a century and placed the territory at the center of an extractive model of capitalism, which resulted in the progressive deterioration of that fictitious commodity, land.[8] It is a research process that aims to create profit but results in the irreversible depletion of the resource itself—which, in an article that has become a classic, Hardin (1968) called "the tragedy of the commons".

Numerous studies on the "shock economy" (Klein 2007; see also Chap. 2) have already underlined how, because of their disruptive and exceptional nature, catastrophic events often constitute opportunities for the promotion of capitalism, opening up unexpected spaces for commodification and value production that look well beyond the event itself, and from which the various "animal spirits" can benefit. The range of possibilities set in motion by the urgency of the crisis include disaster management (according to humanitarian and charitable conventions) (Fassin 2011), followed by the myth of "reconstruction", and any subsequent plans, programs, projects, and expenditure, which, given the exceptional nature of the crisis, are always placed over and above ordinary legislation, thus allowing greater margins for maneuver (and therefore larger gains for those working within the distribution network).

From a practical and material point of view, an "earthquake" implies the need for new forms of spatial and social hierarchization, and with these, different potential opportunities for the development of both economic gain and relational powers (linked to the ability to deal with the intermediation that comes with recovery projects).

The process of building the barracks—technical artifacts doomed over the years to become "shanties", that is, dilapidated housing for the poor—thus goes beyond the "earthquake" and becomes something intimately linked to a model of territorial intervention supported by the rationalizing metaphor of reconstruction and the expression of an economy based on building speculation, on construction as the main, if not only industry present in the territory (see Chap. 4).

Focusing on the contemporary situation involves analyzing the urban layout of the current city as a layered mixture of urban fabric, where new shacks have been added to old ones, in old and new poorer-class areas, or attached to single self-builds or private housing for the middle and lower-middle classes. As is seen in more detail in Chap. 5, it is into this mix that the various territorial stakeholders are ready to move in to play their hands, in particular, the political-administrative class, building contractors, and related industries. Marginal groups that aspire to social mobility by means of work or housing are prepared to make even greater efforts, even in an opportunistic direction, through the logic of rational calculation (showing that, far from representing modernity, they are closer to a utilitarian paradigm of action).

In analytical terms, the intertwining of structure (the expansion of capitalism in local societies, demonstrated by the entry of the State and market) and conjuncture (the earthquake) along which history unfolds (Braudel 1982) can help us to recognize that this is not an incomplete or distorted modernity. On the one hand, the earthquake created an even wider fracture as it "took place during a moment of transition, when certain dynamics were already in progress in communities, speeding up the process of modernization, and obviously taking modernization itself in a certain direction" (Gribaudi 2010, p. 92). On the other hand, it profoundly affected the future of modernization in Messina, as a consequence of the earthquake itself and, subsequently, the powerful metaphor of "reconstruction", which helped to outline the boundaries and the way capitalism developed in the city over the next one hundred years.

In his essay, Faeta (2009, p. 147) invites us to consider the idea that there is "no South and no North", and re-positions contemporary analysis in "the factual immanence of the present" rather than the "nebulous transcendence of the past" (Faeta 2009, p. 161). This interesting image might be a useful way to interpret the Messina case. In fact, the narratives regarding the earthquake and post-earthquake city belong to the same ahistorical past used to observe the South and the Southern Question. This latter is an expression of unfinished, incomplete modernity and the result of the cultural reinforcement of a local population "forced" to live in slums, so that their culture becomes fixed and internalized. If we want to analyze the factual immanence of the present, then we need to do so in a nonessentialist way in order to "grasp the political mechanisms that lead different social actors, at specific moments, to imagine naturalized, essential dimensions of social behavior that are 'ahistorical', 'authentic', and 'anthropological'" (Palumbo 2001, p. 120), suggesting that we "closely investigate the political spaces of production, construction and 'identity'. This means analyzing the ways in which memory and the representation of history are manipulated and constructed, and it also means studying the mechanisms involved in the essentialization and strengthening of the similarities and differences" (Ivi, p. 121).

By focusing the analysis on the contemporary situation and on Messina as a city with "no South and no North", it is possible to investigate the material history of the city and the spatial and social hierarchization of the urban fabric, and how they give rise to a form of modernization that is essentially based on an extractive-speculative capitalism, which revolves around territorial consumption, construction, and territorial management. This model is both entrepreneurial and political (in other words, consensus management) which, far from expressing forms of backwardness, shows a territorially specific way of organizing contemporary capitalism within the city itself.

3.2 A Southern Question? Beyond the Myth of Development

In the first part of this chapter, the aim was to underline the importance of reading the city of Messina and "its" earthquake as integral parts of the narrative regarding the South and the backwardness and underdevelopment implicit in the Southern Question. This interpretation connects the many "Souths" of this world and the way they are imagined and represented. The idea of a "South" which is negatively reflected in a "North" has existed and been reinforced by the symbolic geography of development as the organizing myth of Western "discourse".[9] Development discourse implicates a geopolitical space, modernity, in which the "discovery" of development implies, by negation, underdevelopment (Esteva 1992, p. 7; Escobar 1995, p. 6), and where history is seen as a forward-moving, one-directional, and simple progression toward "modernization" and "progress" (see among others: Gunder Frank 1972; Germani 1973; Schneider and Schneider 1976; Hilowitz 1976; Abu-Lughod 1989; Shiva 1989; McMichael 2004; Leys 2005). Essentially, modernization would resolve a series of opposing dichotomies (which change over time, but are essentially "redundant"): community/society, country/city, tradition/modernity, local village/national state, particularism/universalism, ascription/acquisition, nature/culture, local/global, backwardness/civilization, magical beliefs/rational beliefs, south/north, east/west, and First World/Third World (Antweiler 1993; Agrawal 1995; Dei et al. 2000; Davis 2006a). Here, we are dealing with a stereotypical and essentialized "discourse", but one that is presented as "scientific" and "objective", and which produces specific hierarchical effects, and establishes a discursive order that has the effect of disciplining specific power practices, effectively legitimizing them (Foucault 1997). In this sense, the "South" is seen as an imagined community (Anderson 1991; see also Gruzinski 1988) that has "needs and problems, but few choices and no freedom to act" (Escobar 1995, p. 8) and which, therefore, can and must be the object of "civilization" and "development" policies.

As Escobar suggests (1995, p. 9 passim), in order to reveal these objectifying mechanisms which are the basis of subalternization, the researcher's task is to analyze "developmentalism as a discursive field" and the "modern economic practices" it produces, in the full awareness that the "production of discourse under conditions of unequal power is what Mohanty and others refer to as 'the colonialist move'. This move entails specific constructions of the colonial/Third World subject in/through discourse in ways that allow the exercise of power over it". Here, we might find useful Bhabha's (1990, p. 75) definition of "colonial discourse" as "an apparatus that turns on the recognition and disavowal of racial/cultural/historical differences. Its predominant strategic function is the creation of a space for a 'subject peoples' through the production of knowledges in terms of which surveillance is exercised and a complex form of pleasure/unpleasure is incited [...] The objective of colonial discourse is to construe the colonized as a population of degenerate types on the basis of racial origin, in order to justify conquest and to establish systems of administration and instruction [...] I am referring to a form of governmentality that in marking out a 'subject nation' appropriates, directs and dominates its various spheres of activity".

In relation to the critical and post-colonial literature which criticizes the power relations and spatial and social subalternization present in Western development narratives, Jane Schneider endeavored to translate these problems into the Italian context, and she began the introduction to her book *Italy's 'Southern Question'. Orientalism in one country*, by underlining how:

> In Italy, and in Italian studies, the "Southern Question" evokes a powerful image of the provinces south of Rome as different from the rest of the peninsula, above all for their historic poverty and economic underdevelopment, their engagement in a clientelistic style of politics, and their cultural support for patriarchal gender relations and for various manifestations of organized crime. This tenacious catalogue of stereotypes includes, as well, the notion that southerners, by dint of their very essence, or at least their age-old culture and traditions, possess character traits that are opposite to the trait of northerners: passionate, undisciplined, rebellious, intensely competitive, and incapable of generating groups of solidarity or engaging in collective action, they were and are, as the cliché would have it, unable to build the rational, orderly, civic cultures that, in the North, underwrote the emergence of industrial capitalist society. (Schneider 1998a, p. 3)

Schneider (1998a, p. 4 passim) herself observed that this type of stereotyping resulted from a "Europe-wide 'popular cosmology' of North-South oppositions, in which both northerners and southerners see the 'other' as different and in some respects morally inferior". In other words, it was part of an essentialist dichotomy which functioned as a "key ordering principle of political inequality" (Herzfeld 2005, p. 15), imposing "a model of cultural superiority" within a "particularly powerful instance of the new geopolitics of civilization" (Herzfeld 2005, p. 16).

With her provocative argument that Italy's Southern Question generates a sort of "orientalism in one country", Schneider was calling for the de-construction of the Southern category.

In truth, this process was already underway in Italy. During the 1970s, several scholars had highlighted the fact that the history of the South with its North-development and South-underdevelopment dichotomy had become part of the Southern Question itself. They also showed how its history was an ideological trap, as it did not allow for the inclusion of the Southern regions in the analysis of national and global capitalist dynamics (Mottura and Pugliese 1975; Bagnasco 1977; Bagnasco et al. 1978). More or less at the same time, other studies were beginning to demonstrate how diverse the South was in terms of cultural and social landscape construction (Rossi-Doria 1969; Inea 1976; Sereni 1972; Bevilacqua 1989).

In the same vein, during the second half of the 1980s, a series of books on the history of the Southern regions was published by Einaudi, and the IMES, Istituto Meridionale di Storia e Scienze Sociali (The Southern Institute of History and Social Sciences), was founded along with its accompanying journal *Meridiana*. These two events encouraged a reconsideration of the Italian *Mezzogiorno* and the way it had been instrumentalized (Morris 1997, p. 13). On the one hand, the view of the *Mezzogiorno* was broadened, and the history of the South separated from the Southern Question so that it could take its place in the wider history of European capitalism. There was a call to see the Southern territories not as an exception, but as just another fragment of modernity, in the context of the new urban and rural elites that emerged between the nineteenth and twentieth centuries, and there was a revolution in the capitalistic unification of the market and the consumer society (Lupo 2015). On the other hand, however, the point of view became narrower and the idea of abolishing the *Mezzogiorno* (Viesti 2003) as a single category was put forward, and it became necessary to look at the micro-contexts within the territory, focusing on how they interact with global capitalism. It is noticeable how, even in the peripheral contexts of the world economy, such as

Calabria, for example, capitalism can take root, diversify, and adapt to local institutional structures in a multiplicity of ways and become part of extremely diverse social structures (Arrighi and Piselli 1987).

Similarly, there was a long series of historical studies which proposed studying the *Mezzogiorno* as a concept and a "social construction" (Petrusewicz 1998b, p. 28). "How did the 'Mezzogiorno' become a question?" Marta Petrusewicz (1998a) asks rhetorically, and "How and when did southern Italy become 'the South', a place and people imagined as being different from and inferior to the rest of the country?" Moe asks (2002, p. 1) in the introduction to his book *The View from Vesuvius: Italian Culture and the Southern Question*.

As Dickie puts it (1999, p. 148), "An important part of the answer to these questions about national identity in the Liberal period, and throughout Italy's post-Unification history is 'the South', that space 'beyond' Italy, which is also taken to betoken the whole country's failure to live up to the standards putatively set by nationhood, Europe, or modernity".

In order to answer these questions, we should reiterate that the *Mezzogiorno* does not represent an analytical category and is not a uniform physical and geographical space[10]; instead, we can see it as the expression of a symbolic and political geography that became more solid within the framework of the Italian Unification and the rather uncertain plan to create a modern unified, capitalistic state.

A series of historical studies show that in the aftermath of the Italian Unification, the real goal was not bureaucratic and administrative unification, but rather a practical, ethical-moral, and political question: how to "create Italianness". The solution to this problem required specific positioning and some field choices that became ideological premises. The positioning was positivistic, showing the need for unilineal and progressive development as the only possible way forward. On one side, the field choices implied an essentializing dichotomy between "us/them", and North and South, and the South (which until that time had not existed as a unitary region). On the other, it was necessary to include these dichotomies in development discourse in an attempt to bring order between the two sides, which culminated in the idea that the Savoy North was the desirable model for the national state (Gribaudi 1997).

It becomes evident that the construction of the nation state, trapped in a positivistic and progressive vision of unilineal development, was constrained by the necessity to invent/distinguish between the *Mezzogiorno* and its other, or "alter" (Teti 2013), thus giving rise to the Southern

Question. As Petrusewicz puts it (1998a, pp. 9–10), the Southern Question creates "a relationship of reciprocal otherness between the North and the South, both ideal types: in order to exist the Southern Question requires an Italian context, real or imaginary, [...] the South and its 'alter', the North. When we refer to the 'South', historically we are talking about the Kingdom of the Two Sicilies, while the 'North' was the Kingdom of Savoy. Where the 'South' is synonymous with a peasant culture, the 'North' is urbanized and industrial, and if the 'South' is the producer of agricultural foodstuffs, raw materials and the provider of cheap labor, the 'North' represents manufacturers, carmakers, specialized services, expertise and workers represented by Trade Unions. Where the 'South' is a Mediterranean civilization based on Greek and Arab traditions, it contrasts with the 'North', which is Central-Western European and Lombard. Since they are imaginary constructions, there is not a great deal of sense in making objections or verifications: South and North exist only in relation to this 'otherness', which presupposes a pairing and an imaginative complicity".

By inference, the South became everything that the North was not, and it was identified with a stereotypical "Italianness" (Gribaudi 1997).

To avoid this type of reductivist impasse, it was necessary to detach the South from the Southern Question (Bevilacqua 1993; Lupo 1998), to emphasize the role of the intellectual elite and southern politics, to stop producing and reproducing the *Mezzogiorno* as the "Southern Question" (Lupo 1998, 2015), and to detach the South from meridionalism (Giarrizzo 1992).

Meridionalism is a school of thought defined as a "substantially consistent and concordant tradition of analysis and debate on the Southern Question, and can be seen as a movement for the defense of Southern Italy" (Lupo 1998, p. 19). In the end, the term becomes an umbrella category, uniting authors, analysts, and thinkers belonging to profoundly different eras, by the idea that a Southern Question does exist, but that they find very different, often contrasting, explanations and solutions for it, which sometimes descend into cultural or moralistic notions about the people of the South (Giarrizzo 1992; Lupo 1998).

In this respect, the intellectuals contributed to strengthening the representation of the *Mezzogiorno* as the Southern Question (by underlining the differences between North and South), and legitimizing the school of thought. If we look at the large number of essays, pamphlets, and investigations that were written from the time of the Italian Unification to the end of the Second World War, and which formed the basis for the debate on the "Southern Question", we can see that the political connotations were mixed and even conflicting. It is evident that we are "talking about a Southern identity and about how the intellectual classes in the South reacted to the unitary State, both positively and negatively, and how the leaders of the new state regarded Southern Italian civilization" (Placanica 1998, p. 159).[11]

The economists play a central role in this group. Throughout this chapter, we have highlighted how the mechanisms responsible for creating the Southern Question in the Italian *Mezzogiorno* were the result of economic underdevelopment and cultural variables and differences. Traditionally, economists have "measured" the Southern Question as a divergence between the main statistical indicators that detect the level of economic and social growth (with the GDP as the main indicator, followed by other structural variables such as the rate of labor market participation, education levels, levels of poverty, and so on). Development is therefore a process of growth that initiates convergence, helps to bridge the gap, and brings this part of the country "back" to the same levels of social and economic well-being as the rest of the nation. A large part of the recent economic debate about the South, therefore, has been about "measuring" the process of convergence in terms of territorial differences, disparity in the GDP, and other economic indicators, stimulated by the discovery that since the 1990s, this process of convergence, which began after the Second World War, has suddenly been interrupted. The diachronic analysis of certain historical periods have been carried out (Malanima 2003; Daniele and Malanima 2007, 2014; Malanima and Felice 2013; Felice and Vecchi 2015; Felice and Vasta 2015) in order to identify precisely when these developmental differences occurred and in which phases they were seen to increase and decrease. In short, the analysis shows that the gap in the per capita GDP was practically non-existent in the period immediately after the Italian Unification, nor was it so clear-cut in the first few decades after that; the differences only became more apparent during the second half of the nineteenth century (1880) and up to the Great War, as a result of industrialization policies. A further increase took place during the Fascist period, as the policies of this regime tended to reinforce the "industrial triangle" (i.e., the richest and most industrialized part of Italy, which includes the northwestern areas of Turin, Milan, and Genoa). These reached their peak in 1951, when two decades of extraordinary convergence began, driven by both the interventions of the Fund for the South

(*Cassa per il Mezzogiorno*) and the economic boom. By the 1970s, the convergence was already beginning to decrease, due to the energy and oil crisis and changes in the model of industrial capitalism that made primary and intensive industry obsolete, something that had been central in driving Southern development. From the early 1970s onward, a progressive reduction in the rates of convergence began, leading to the crystallization of the gap between the Center-North and South: the southern regions were unable to reach 60% of the income of the Center-North regions. A further increase in the gap was seen at the beginning of the 1990s, and it has continued up to the present. This was also a consequence of the European Union's (EU) austerity policies to contain the public debt and public spending, which had a deflationary impact on aggregate demand and did not increase labor productivity (Daniele and Malanima 2007, 2014; Farinella and Ofria 2016).

Although it is interesting to try to quantify "southern backwardness" by measuring the level of disparity and analyze its duration, this operation can be problematic and dangerous.

It can be dangerous because it only indirectly confirms the irreducibility of the gap: if, after 150 years of debating the Southern Question and policies that might help "resolve" it, and despite brief intervals ("Les Trente Glorieuses" or economic miracle, as it was called, respectively, in France and Italy, with reference to the period from 1945 to 1975), nothing has substantially changed. So what else can be done? Is this gap caused by cultural factors that are so deep-rooted they can never be changed? This is the sort of reasoning that can result.

It can be problematic because the GDP mentioned in these studies is always a preconceived starting point. In fact, the real utility/ability of bringing to light the supposed "Southern Question" is not in doubt, especially if we consider the diachronic plane. One wonders whether an indicator created to "measure" "the wealth of the population" today, "is able to grasp, using questionable statistical data from the late 1800s, the differences between local economies in remote historical contexts that are so different from the ones for which it was devised" (Lupo 2015, p. 4). Moreover, international debate has often underlined the inadequacy of using the GDP to measure wealth (Stiglitz et al. 2009) because economic development and social growth do not necessarily go hand in hand.

In order to respond to these criticisms, economic literature has complicated the statistical analysis by including variables of a more socio-institutional nature, which should reveal some of the more intangible components of social structure, such as the level of social capital (considered an indicator of the ability to trust and cooperate for growth) or the level of institutional efficiency (the ability of the institutions to guarantee compliance with the rules), so-called civicness, the degree and type of "familism" present, on the basis of a series of neo-institutionalist and culturalist analyses which began with the studies of Banfield (1958) and Putnam (1993).

In 1958, Banfield published the results of a case study he had carried out in Montegrano, a small community in Southern Italy, in a book evocatively entitled: *The Moral basis of a Backward Society*. The book revolves around the theory that the backwardness of the local community (where the villagers are mostly described as lacking in initiative, indolent, and lazy) was the result of a concept that he called "amoral familism". According to this theory, each nuclear family is exclusively interested in guaranteeing the interests of its own members in an opportunistic and individualistic way, to the detriment of the community. This absolute reliance on the family and its material needs reveals how, in a context of scarcity, the situation can degenerate not only into an inability to trust and cooperate for the production of collective goods but also in their appropriation for personal use, which translates into widespread backwardness. Amoral familism, therefore, is the tendency toward a form of ethics based on selfishness (*self-regarding*) which promotes the well-being of the inner circle (family, relatives, friends) to the detriment of collectivity (Piattoni 1998).

In the same vein, Putnam published his famous book, *Making Democracy Work: Civic Traditions in Modern Italy*, in 1993, in which he aimed to give a detailed historical and statistical analysis that would explain the differences in the development of certain Italian regions. According to Putnam, the differences in the development of the two Italian macro-regions could be explained by the different historical evolution of their civic traditions (*civicness*) and by their relationship to the public institutions, which had already begun to take shape during the Middle Ages. On the one hand, there was the central-vertical regime of the Norman state in the South, as opposed to the municipal-horizontal regime of local authorities in the Center-North. Subsequently, this would lead to the consolidation of two profoundly different administrative and politico-social backgrounds, which came about at the time of the Unification. In the Center-North, this translated into a tradition geared toward cooperation, generalized reciprocity, social capital, and the good governance of the institutions. In contrast, in the South, it translated into a non-cooperative tradition with restricted reciprocity and a lack of faith in the institutions.

Both of these studies proposed a cultural interpretation of economic development, according to which it derives from institutional factors (variously, family structure, property regimes, so-called social capital and trust, norms and interior values, tacit knowledge, etc.) which become established within a given territory over the course of time, through specific institutional practices which generate path dependency. These stereotypical explanations of economic inequality reflected widespread modernist ideologies entrenched in development programs both internal to the Global North and imposed by the North on the South, emphasizing regions with a "culture of poverty" to be somehow repaired and replaced (Lewis 1959).

If, initially, the tendency is to concentrate on institutional factors in order to avoid a merely economic approach, it soon becomes obvious that this is a culturalist trap. Carrying out a complex historical causal analysis of the past can help to "account for" current differences in development levels, but there is the risk of being drawn into a culturalist, determinist, and prescriptive drift. In this way, we could account for the existence of territories "condemned" to underdevelopment, territories that are "imprisoned" in their own cognitive, inherited, and socially learned cells. There are numerous difficulties with this type of approach, and with the methodological decisions taken along the way, and there is a high probability that some of these will be completely arbitrary and erroneous. The first difficulty concerns the classic problem of methodology for the social sciences: how to translate abstract concepts into sets of variables? The second involves the impossibility of measuring these so-called institutional elements ("civicness" and "social capital") without running the risk of reifying them. The third problem regards the identification of causal links, also in relation to the researcher's own values, and the "history" that might explain why, in certain territories, some institutional characteristics have become established rather than others. The risk of underestimating the importance of other unforeseen or unconsidered variables is directly proportional to the extent of the period under investigation (the further back in time, the more imprecise the facts become). According to Weber (1906, p. 115; Weber 1907, p. 122), "an infinity of causal factors have influenced the occurrence of the particular event" and the "selection or identification of the object of empirical explanation is 'determined' by its relation to values". The fourth refers to the importance of not absolutizing explanations: as Braudel (1997) pointed out, the importance of a historical fact should always be considered provisionally, so that if need be, it can be revised later. Moreover, outside the orders of discourse that claim they exist, "local cultures" and "historical roots" do not exist now and "have never existed in the past" (Palumbo 2001).

3.3 Is an Agricultural Question at the Root of the Southern Question?

Thus far, the aim has been to emphasize how the representation of the *Mezzogiorno* as the "Southern Question" is related to an essentializing narrative that contrasts the development and certain features and characteristics of the North and South of Italy, which should be universal and mutually exclusive.

The purpose of this section is to underline how the "Southern Question" discourse derives from issues involving the relationship between state, market, and local society. Therefore, the history of the national territory should be seen in the context of the construction of the nation state but also in the context of the global capitalistic economy. When and how this process took place is the subject of extensive national and international discussion (Lumley and Morris 1997; Davis 1998).

Shortly after the Unification, the more liberal historians underlined the fact that Southern Italy's underdevelopment was probably due to the pre-Unification choices made by the Bourbon dynasty. In no particular order, these included the closing down of southern industry, the failure to apply capitalistic modernization to agriculture in areas where, as a result, agriculture became non-productive, there were absentee landowners, and an administrative structure that was riddled with nepotism, inefficiency, and a lack of productivity.

By contrast, other historians have overturned this vision, claiming that the Bourbons had created a modern and advanced socio-economic structure that was literally swept away by post-Unification political choices which tended to favor the North, making the socio-economic structure of the south subordinate, and generating a sort of internal colonialism that produced the South's dependence on the North (see among others, Zitara 1971, 2011). Once again, we are in the presence of a mechanistically dichotomous theory which: (1) overestimates and distorts the role of certain factors (in particular, the importance of the industrial facilities of the period. Lupo 2015). (2) Fails to recognize the importance of state and market deployment mechanisms within the southern territories and the

South's response and adaptation. (3) Obscures some of the post-Unification dynamics, which, although they were not strictly southern, should be considered as part of the more general Italian development model that came into being after the Second World War, for example, the dualistic structure of the labor market. This structure was between *insiders* (adult male workers employed in large-scale industry or the public service sector in a "primary" labor market where employment was guaranteed and protected both from the risk of job loss, thanks to unemployment benefits, and the right to a retirement pension) and *outsiders* (young people, women, and unskilled workers with informal and/or precarious jobs in the secondary labor market, working for small manufacturing, construction, or commercial businesses, where there was no recognition of employment rights such as paid sick leave, unemployment benefits, or a pension) (Mottura and Pugliese 1975; Mingione 1991).

Assuredly more dynamic is the Marxist analysis, which underlined how "the underdevelopment of the South was not an expression of 'historical delay' but rather the birth of the Italian capitalist State" (Mottura and Pugliese 1975, p. 158). In his analysis of the Italian Risorgimento, Gramsci (1975, p. 2011) argued that the Italian Unification, imposed from above and without popular participation, was the outcome of a failed revolution ("passive revolution"): the *Partito d'Azione* had been unable to extend consensus to the poorer classes, especially to the peasantry. The construction of the Italian State was managed and controlled by an elitist bourgeoisie who were unable to lay the foundations for an enlarged democracy. The new state arose from a political socio-economic base that was restricted, oligarchic, and unable to develop a project of inclusive social change. The lack of a solid and broad base of consensus for the Unification project and its subordination to the Savoy Monarchy made the liberal ruling class weak and susceptible to collusive and clientelistic practices in the construction and maintenance of power (Turiello 1974; Silverman 1974). From an elitist concept of the liberal state, a "historical bloc" emerged, that is, a political alliance between reactionary southern landlords and the northern industrial bourgeoisie, which inhibited the modernization of the South: the policies put forward by the historical bloc favored the landowners to the detriment of the peasants and diverted capital from the South to the North, thus weakening the southern economy (Gramsci 1966).

From the post-Unification period up to the years following the Second World War, the Southern Question was closely linked to an "agricultural" question. As Sereni puts it (1977, p. 43 passim), during the early post-

Unification years, the liberal policies of the historical right were disastrous for the poorest peasant farmers in the South. These policies were designed to centralize and standardize all political and administrative systems to the Piedmont model, also extending its legislation and fiscal systems to the annexed regions (Saraceno 2005). The cost of Unification had brought about a substantial rise in the Piedmont public debt, which was inherited by the newly emerging state, generating huge fiscal pressure, and sustained by new taxes on agriculture and land rents rather than securities. In this way, it was the southern peasants who were most disadvantaged, as they were the ones who assumed the costs of land taxation, both directly (due to the increase in the cost of renting or owning small areas of land which they cultivated in order to supplement the family income) and indirectly (with a reduction in the wages they received as farm workers for rich landowners) (Sereni 1977). The peasants' lives were made even more difficult by the abolition of collective rights and the closure of publicly owned land: on the one hand, this meant that the peasants no longer had an important economic resource to supplement their daily lives (both grazing and cultivation rights, and the right to forage and collect wood) and, on the other, the division of publicly owned land ended up benefitting the landowners and landed aristocracy who often closed public land illegally and, through their control of local public administration, often managed to win public auctions (Sereni 1977). This was one of the reasons why groups of brigands formed in the southern countryside, although they were ferociously repressed.

For Sereni (1977, p. 68), the policies of the Historical Right did not affect the entire population of the South but mostly the poorer members, and they were therefore specifically class-related: the dominant classes in the South used bureaucratic centralization and fiscal Unification to maintain dominion, and although the Monarchy was forced to ally itself with a small group of southern gentlemen, it managed to keep them in a position of relative subordination. It is important to recall here that, according to Pantaleoni (1891, p. 68 passim), for the period between 1884 and 1889, northern Italy had a per capita average income of 2411 lire, owned 56% of the national wealth, and paid only 47% in state taxes, while the South, with a per capita income of 1372 lire, owned 27% of the national wealth and paid 32% of the taxes.

Moreover, right-wing liberal policies, which aimed to break down the customs barriers and intensify trade, flooded the Italian market with low-cost foreign products, separating industry from agriculture, causing a decline in those small cottage industries that were so important to the production and reproduction strategies of rural families. The new national market was increasingly a market for large-scale industries and, as such, the only one able to produce at a low cost. The liberalization of the customs system also stimulated the commercialization of the Italian economy, so that for the first time it was exposed to the volatility of international market prices, and it was necessary to resort to what Sereni (1974, p. 94) called the industrial market's "artillery of low prices", promoting the development of agriculture in a commercial and capitalistic sense.

While the disappearance of cottage industries in the North was offset by the growth of large-scale industry, the South "was subject to a process of agrarianization, and the bulk of the labor that the peasant population had previously dedicated to industrial work, was left unused" (Sereni 1977, p. 39). And it was only a little later on, when a protectionist regime had been established to oppose the entry of foreign goods, that the Northern industries realized that the South was a new growing market, to be used in an almost monopolistic manner (Sereni 1977, p. 38): the construction of a national capitalist market thus sanctioned the creation of the Southern Question in which the industrial North was always in opposition to the agricultural South.

It is important to note that right-wing liberal economic policies had helped to promote new ideas for some segments of southern agriculture, encouraging the development of a new intensive specialized form of agriculture (olive oil, vines, citrus production) destined for the international market. In 1887, the so-called Historical Left came to power, bringing in a period of political protectionism that favored the steel, textile, and grain industries. This was the result of a hiatus between the industrial interests of the North and the agricultural interests of the southern landowners. However, this measure undermined the only dynamic sector of southern agriculture: specialized crops. The pressure and discontent felt by the increasingly proletarianized peasant masses, slowly weakened during the Crispi and Giolitti governments, and was one of the reasons behind their emigration to the new African colonies and the USA, which acted as a relief valve for the subaltern classes.

In this sense, the long history of Sicily, from the silk and wheat production of the late 1700s to the intensive cultivation of citrus fruit in the mid-1800s, is a very interesting way of trying to understand some of the mechanisms set in motion at a time when the southern regions were being incorporated into the developing nation state. This process was also part of the wider dynamic of the capitalist market.

Far from describing an image of pre-capitalist Sicily with a local subsistence economy that was impervious to the outside world, Braudel (1982) had already emphasized how sixteenth-century Sicily and the Mediterranean area were characterized by a rich economic life that took place on three different interconnected levels, and where manufacture, products, and the way they were used could be diversified. On the one hand, there were long capitalist supply chains of international trade, reserved for specific products, and suitable for long-distance exchange and particularly sought-after abroad. According to Braudel (1985), these are the supply chains that constitute the essence of capitalism, that is, a "private market" driven by competition, but which is actually an oligopoly, dominated by just a few subjects who control the long and consequently rather opaque exchange networks. On the other hand, the medium-length and short-range supply chains, characterized by a market economy, where trade was mostly oriented toward free competition and the public market, with numerous participants, mostly small traders. Finally, the great substratum of material life oriented toward the reproduction of local subsistence economies, within relatively closed and protected economic circuits, in which monetary exchange is not as important as it is for the other two.

Braudel (1982, 1985) also underlined the fact that there were numerous points of contact and interaction between the different economic forms, for example, short-range markets or fairs, which could connect traditional trade with long-range supply chains. In this respect, various studies have shown that, in the case of Sicily, changes to the agricultural landscape for the cultivation of different crops were not casual; on the contrary, they were the result of rational strategies, which demonstrate how these crops were dependent on the international market, and the long global capitalist chains to which they were precariously linked. Even though this connection to the international market was one of subordination, it allowed producers to "reinvent" cultivation and production methods based on demand, and in a low-grade speculative way as far as international competition was concerned (poor technological improvement, low wages, high intensification, and movement to different sectors to gain higher profits). However, this would prove to be fatal in the long term.[12] As Laudani points out (1989, p. 143), from the Sicilian experience, we can see there was a tendency to abandon a production sector at the first signs of profit erosion or a reduction in demand, but thanks to the high level of agricultural specialization, it could keep up with international markets by turning to alternative products. "These are the contradictions of 'a rich agriculture in an underdeveloped area' [...], which closes one production phase to open another, does not make the most of its capabilities, does not make investments or technological improvements to become truly competitive, but rather, recognizes the fertility of its soil, its low wages and insularity as its greatest opportunities. [...] From sugar cane to silk, from silk to wine and citrus fruits (not to mention Sulphur mining in the Sicilian interior, as well as wheat and agricultural estates), the Sicilian economy managed to reinvent its productive identity numerous times".[13]

For example, the production of silk and silkworm farming, which was so important to the local economy in Messina, was a cottage industry operated by peasant families. Connected to this were the cultivation of the mulberry tree for the silkworms to feed on, and spinning and textile production, which were carried out either in the peasant families' homes or in small workshops. Thus, agriculture and manufacture came together as complementary aspects of a home industry. For these peasant families, silkworm farming was seen as an activity that would help supplement the family income and was generally carried out by women. Silk was produced for the national and international market rather than for the peasants' own consumption, and cities such as Messina were an important part of the supply chain, not so much as market outlets, but as centers for the organization of business and trade. This was also thanks to the fact that they were large commercial ports, used as storage and exchange hubs. Lupo (1987, p. 91) highlights the fact that already, in the early modern period, Messina's physical characteristics were not adapted to monocropping grain, but had brought about "the cultivation of woodland and precociously diversified agriculture. Along with the ubiquitous vineyard, the cultivation of mulberry trees was widespread as the leaves were essential to silk production; and it was thanks to silk production, whether raw or (more rarely) processed, that Messina had modelled its economy and institutions so that it 'almost gave the impression of being a republic' within the Spanish empire. It had a sophisticated commercial and credit structure, based on the merchants' willingness to advance payment to producers and on the small network of cottage weavers, which gave the proud city of Messina a dominant role embodied in the monopoly of the island's silk exports".

Agriculture, industry, and commerce linked together to form a capitalist economy, selling for the market. But it was the mechanisms of capitalism itself that initially caused the crisis in production when the international scene changed: the mechanization of the weaving industry that was tied up with the emergence of Northern European, and, more specifically, English large-scale industry, determined the arrival of low-cost foreign yarns which put local non-mechanized products "out of business" (Laudani 1989; Sereni 1977).

Commercial cities such as Messina were losing some of the production power that had stimulated their growth in the first place, so they tried using different resistance strategies. On the one hand, they rode the wave of commercialization, emphasizing their role as a hub for the exchange of goods, and on the other, they tried modifying their production and commercial framework, changing to alternative products that were in growing demand on the international market. This is precisely what happened with the growth of intensive cultivations of olives, vines, and citrus fruit—particularly lemons and oranges (among others, Lupo 1984). The demand for these products, which had already been high during the pre-Unification period, increased with the opening up of international markets in the post-Unification period, and replaced other cultivations (wheat and mulberry trees) that were going through a period of crisis stimulated by the same market. It was during this period that the agricultural landscape in Sicily was organized to look more like a "garden" (Sereni 1972; Lupo 1984, 1987). The intensive and specialized cultivation of citrus fruits on irrigated land was mainly destined for exportation, and connected to global supply chains as well as local trade. Within this framework, Messina was, yet again, an interesting case, as is examined in more depth in Chap. 4. The city managed to organize its agricultural production in relation to demand and international trade, thus keeping its commercial orientation viable. This commercial focus is evident from the diversification and de-seasonalization of products, for example, the city was specialized in the production of "verdelli", or "lemons that were forced to fruit in spring in order to satisfy the summer market. [...] The 'forcing' technique for the production of 'verdelli' cannot be repeated every year in a single lemon orchard, and it is necessary to use the normal winter fruit. The producers therefore adopted a rational strategy, placing all their care and investment in the 'verdelli', and allocating the rest of their produce to the derivatives [e.g., juice] industry, in order to save on the costs of cultivating and harvesting the fresh fruit. To sum up, Messina produced the highly valued early lemon, as well as a more inferior variety, while only the medium quality winter fruit was missing, and in this way the city managed to consolidate its central role in the derivatives industry" (Lupo 1987, p. 93).

Moreover, there was a solid system of exchange and interconnection between the city and the country, with the city as the center of commerce and industry. Meanwhile, the peripheral villages cultivated produce for the Messina marketplace and, during the winter months, provided peasant workers who were able to supplement their family income by working as manual laborers in small-scale industry and construction.

Connected to this system of networks between city and country, there were different intermediaries who guaranteed continuity. The volume about Sicily (Lorenzoni 1910, p. 450), the *Parliamentary Inquiry into Conditions in the Southern Provinces and in Sicily, 1909–1910*, showed that it was highly urbanized and organized into "huge rural villages", the so-called agrotowns,[14] which, for various reasons appeared to function: (1) as a place of residence for the peasants, and from where they could control their itinerant work and the breaking up of property (Rossi-Doria 1948, p. 16). (2) As a marketplace for daily and seasonal laborers to find work (Bevilacqua 1984, p. 54). (3) As a "production service structure which [...] drawing together all the agricultural work not directly carried out in the fields, [and where] the long-distance trade terminals met, and the medium-length terminals were organized so as to connect the various segments of highly specialized cultivation" (Salvemini 1989, p. 116).

The cities of Messina and Palermo can also be added to the list of rural towns. By 1861, they already had over 100,000 inhabitants (Barone 1989), making them second only to Naples in size in southern Italy. Far from being "tertiary-agricultural hubs" designed to accommodate offices, these cities had a more complex function. The port cities such as Messina, for example, with their direct links to rail and road infrastructure networks, were not only locations for the storage and transit of goods, but they also had a series of industrial and commercial activities. As such, they became "places for the transformation of products and for the organization of commercial services for the «rich» countryside [...] In Catania, at the end of the nineteenth century, the chimneys from the sulphur refineries and chemical plants (around 70 factories with 2000 employees) surrounded the railway station and docks, while in Palermo, the Oretea foundry and the Florio mechanical workshops ran alongside the shipyards and dry docks. In Messina, the Stinnes coal warehouses and the railway companies maintained the supply of raw materials to Sicily and Calabria" (Barone 1989, p. 33).

These observations show that the southern city also dominated the countryside (Giarrizzo and Maraini 1980), and that it was decisive in making choices for agricultural investment, and responsible for the reorganization of the countryside. It also provided the peasants with alternative or complementary work opportunities not necessarily connected to agriculture. During the 1870s, the "meridionalist" school of thought had erased the role of the city, "favoring the analysis of the countryside, and underestimating the impact of the changes in the political system, which were modifying the balance of power in southern society in both larger and smaller cities" or "claimed that the real roots of the southern question could be traced back to the peasant movement and a supposed and uniform 'rurality'" (Barone 1989, p. 39).

A well-established line of studies concerning "industrial districts",[15] (territorial systems with small and medium-sized artisanal businesses in the Center-North of Italy) has long emphasized the positive role of northern cities on agriculture and their help in creating the necessary synergy for the development of the small manufacturing businesses that are so vital to the contemporary Italian economy. On the contrary, southern cities have been negatively conditioned by the "backwardness" of agriculture and because they have been unable to develop productive systems that rely on consumption and rents. In the Center-North, in particular, different farming methods (share-cropping rather than the "latifondo" that was more common in the South), and other factors such as family structure, helped to produce the necessary manpower and capital for starting production activities, as well as the right entrepreneurial spirit for the creation of industrial districts. In reality, studies have shown that this is a rather ungenerous picture of the South. Petrusewicz (1996), for example, underlines how agricultural estates can be considered dynamic structures, and not simply an expression of backwardness. As Davis puts it (1998, p. 213):

> As a creature of the nineteenth century, and of the combination of high value-added cash crops with the more traditional extensive pasture and wheat-growing, the *latifondo* was commercially oriented, creating a "system" that permitted considerable flexibility in responding to changing market conditions. When export markets were poor, the system could fall back on the extensive sector and its own internal markets. Marta Petrusewicz, for example, stresses how the management of the vast Barracco estates in Calabria constituted a commercially oriented form of production that was rational and made highly flexible use of the resources available through a system that again sought to minimize the risks inherent in those markets.

Other studies show that the *Mezzogiorno* was not simply characterized by an indistinct latifundium, but also by the coexistence of different agricultural systems (among others see Rossi-Doria 1948; Arrighi and Piselli 1987; Sereni 1972; Bevilacqua 1993). Within this framework, there was a growing dynamism between city and country, and the city was increasingly ahead of the country in matters of modernization. "The southern city was not just a passive urban area in comparison to the backwardness of the countryside; during the twentieth century, the process of functional hierarchization gave form and substance to a new spatial hegemony, and strong economic polarity. The towns and villages that produced olive oil, wine and citrus fruit, the Sulphur-mining towns, and the very same 'agrotowns' which, by the end of the century were specializing in the production of durum wheat for the national milling companies, organized themselves as the locus of control for the transformation of the land, and their results were based on the alternating phases of global demand" (Barone 1989, pp. 20–21).

By recognizing the dynamic and complementary nature of the relationship between the city and country, it is possible to underline the different roles played by the peasants (Rossi-Doria 1948; Tarrow 1967; Sereni 1977). The peasant family, which was both a unit of economic production and social reproduction, acted as a "clearing house" (Wolf 1955; Meloni 1984, 1997; Signorelli 1983), in other words as an elastic and ductile productive-reproductive structure. The family was a place for the integration and coordination of various, diversified, and discontinuous incomes, which were maximized. The family also encouraged self-exploitation, the activation, and de-activation of parts of the workforce; helped reduce needs; and created non-monetary income. The total family income was never exclusively dependent on a single source, and the peasant never engaged in one single activity. In fact, they were often small landowners or tenant farmers, but their plots were so small that subsistence was not guaranteed. For this reason, and depending on the alternation of the crops and the different territorial specializations, they often worked as paid seasonal and non-seasonal agricultural laborers on the latifundium (Rossi-Doria 1948; Tarrow 1967). However, they also worked as laborers on the outskirts of the city—as in the case of Messina, where de-seasonalized production freed up a lot of the peasant workforce during the winter months (Lupo 1987)—and undertook "industrial work at home" (Sereni 1977). It was therefore impossible to find a single productive role, and in this way, the peasants adapted to a system of highly territorialized agricultural characteristics. The alternation/integration between agriculture and livestock farming, the widespread presence of mixed crops in some areas, and single intensive mono-cultivations or intensive specialized ones in others, the role of woodland in the Mediterranean maquis, all contributed to the definition of the agricultural landscape which can be seen as a mosaic of cultural, ecological, and historical biodiversity (Sereni 1972; Agnoletti 2013).

Against this background, Giolitti's policies of customs protectionism proved to be fatal for the Italian *Mezzogiorno* because they not only created a crisis in the most dynamic capitalistic sectors of local agriculture but also trickled down to the more traditional sectors, with which they had previously been in equilibrium. At the same time, these policies advanced the capitalist modernization of agriculture, which meant an increase in productivity and a reduction in the workforce, until the advent of Fascism.

The policies of the twenty-year Fascist period had imposed ruralism, and also revealed a strong latent anti-meridionalist sentiment (Barbagallo 1973). While national policies continued to strengthen industry in the North of Italy, the enforced ruralism proved to be a way of condemning southern peasants, once again, to a life of subsistence. Fascism not only strengthened the so-called historical bloc, an alliance between the industrialists in the North and the large landowners in the South, but also included in its organizational structure people and intermediaries working for the mafia, who were given the task of repressing peasant uprisings. One of the most problematic aspects of the transition from Fascism to the new republican state was the fact that this mixed group of southerners (in particular, the large landowners, large business and property owners, mafia bosses, and intermediaries who had become part of the Fascist administration), who had gained strength under the Fascist regime, managed to keep their positions within the new democratic institutions without being purged.

This was made possible by the necessity to guarantee the functionality of the historical bloc during the republican phase, as Davis (1998, p. 207) points out: "not only for Gramsci and the Left, the 'Blocco storico' has widely been seen as the cornerstone of Italian politics from the protective tariffs of the 1880s to Mussolini's agrarian autarchy—indeed, for many, the Christian Democrats' success in turning the South into an electoral fortress after 1949 has been seen as a continuation of the politics of the *Blocco Storico* by other means".

It is important to keep these ideas in mind in order to understand the political choices that were made after the Second World War. According to

Mottura and Pugliese (1975, p. 168 passim), in a framework where agriculture was becoming increasingly secondary to industry, the agricultural policies of the post-Second World War period were designed to stabilize the socio-political system. The law regarding land reform (L. n. 841 del 21/10/1950) was supposed to allow peasants to buy land, but it actually referred to a few specific areas only, and it represented only 40% of national agricultural land (around 680,000 hectares, just over half a million of which were in the South). The reform did not aim to improve conditions for peasants, but to address the unemployment situation for peasants and laborers, and reduce social tensions. It also went some way toward helping to break up the peasant movement, thereby reducing the pressure on the peasant labor market and creating instead a potential army of industrial reserves that could be called on at any time. Furthermore, thanks to land reclamation, the reform also benefited the agrarian capitalists who saw the value of the land near the reclamation areas increase dramatically, and received large amounts of compensation for potential expropriations, thus generating a flow of capital that was mostly transferred to urban real estate through property speculation (Mottura and Pugliese 1975, p. 168).

In general, post-war agricultural policies helped supplant peasant-based agriculture in favor of capitalist agriculture that was more oriented toward low-cost agricultural products; this meant that productivity increased but occupation was reduced, and there was a gradual depopulation of the internal areas where mechanized or intensive agricultural methods could not be used.

In more general terms, the decline in peasant agriculture (where the peasant was perceived as a mixed figure and the peasant farm depended on the ability of the entire family to generate income, including non-monetary income through the practice of multi-functional and multi-activity strategies) can be seen as a progressive "dis-embedding" from the local community and "embedding" in a State and market logic. The emergence of capitalism in agriculture implies the progressive commodification of agricultural labor, so that the peasants lost autonomy and became completely dependent on salaried work, but in a sector in which demand for laborers had diminished sharply, thanks to the modernization of agricultural methods.

Post-war agricultural laborers were the expression of surplus labor, falsely immobilized as underemployed labor in the countryside, part of the newly emerging Fordist social system (Mingione 1991), and a potential reserve army of low-cost manual workers for areas of intense industrial development. However, precisely because of their role as "salaried workers", albeit seasonal ones, and the fact that they were also classed as "unemployed", they were the recipients of the "welfare" policies that were emerging as part of the new care system and the extraordinary intervention for the South. The problem of work and agricultural sub-employment was therefore solved by individual specialization in the search for particularistic-clientelistic channels, in order to gain access to the benefits provided by the modern welfare state, a job or income (pension or social benefits) from the public or para-public sector. They were therefore a potential army for the political consensus of the public institutions, especially as part of a strategy aimed at containing social tension and the growing conflict of the working class in the Center-North. Indeed, this is how we should interpret the decision to establish the "pension for agricultural laborers" or the re-employment of these workers as manual laborers in the construction industries of the South. These industries had managed to gain a foothold, thanks to the infrastructure investments paid for by the extraordinary intervention and to the building boom taking place in many Italian cities (driven by the policies for the construction of social housing for the poorer classes and incentives for buying a first home, and later, for the middle classes, incentives for buying a second home).

According to some classic studies on clientelism, it is at this stage that clientelism among the more notables, which is vertical and typical of rural areas, shifts to a bureaucratic type of clientelism, based in the city and organized in a horizontal way, and where the intermediaries become administrative or party officials (Tarrow 1967; Graziano 1974; Blok 1974; Gribaudi 1980; Li Causi 1976). Despite the importance of underlining how clientelism and intermediation changed after the Second World War in response to a social context in which public spending had become the backbone of the economy, Piattoni (1998, p. 225) believes that this sort of analysis is risky, as it places clientelism in a position of dualistic opposition. In itself, it is a negative and archaic aspect of traditional society, which is "transferred" to modern society, where it changes and adapts to become bureaucratic clientelism, which, over time, produces and reproduces to become a social bloc. Modern clientelism is therefore the result of the transformation of a so-called traditional society (in which relations are prevalently personalistic, polyfunctional, and ascriptive) into a modern society (characterized by universalism, depersonalized bureaucracy, specialization, acquisition), which leads to an imperfect form of modernization.

On the contrary, clientelistic networks are not strictly connected to "traditional" or "backward" businesses and are particularly flexible, placing themselves at the service of contemporary capitalist systems with cliques that change over time, adapting to political and economic trends and conforming to the groups in power or to specific objectives. Clientelism should not be seen as solely negative, but it should be analyzed in a neutral way, as a strategy for collecting votes. These are networks of exchange geared toward producing selective benefits that are advantageous to the few and can, in certain contexts, produce collective benefits either for a wide category or for all (Piattoni 2007, p. VIII). In short, both clientelism and particularism can be positive resources that help to generate trust and can trigger virtuous processes in certain social and institutional conditions, as Mutti (1994) and Piattoni (1998) have already highlighted. For this reason, rather than talking about abstract clientelism, we need to talk about a "clientelistic game", and analyze how this works in real territorial contexts.

3.4 The Fund for the South

The period immediately after the Second World War saw a season of intense debate on the "Southern Question". After the devastation of the war, conditions in the South were particularly challenging, with extreme poverty and semi-literacy. In addition, the gap between the South and the rest of Italy had widened. In an international context, a Keynesian model of social regulation was established, whereby the State "supported" demand through the creation of work and income. In this climate, a "new" group of meridionalist scholars who were part of the growing Association for the Development of Industry in the South (SVIMEZ), realized that the Southern Question needed a centralized political response. The only way to solve the age-old problem of inadequate infrastructure and lack of business development in the South was through robust and continuous public intervention. In 1950, an "extraordinary intervention" was organized for the development of the South, and a public organization was set up to create it, the *Cassa per il Mezzogiorno* (the Fund for the South), influenced by the Tennessee Valley Authority.[16]

The initial plan lasted ten years (and was subsequently extended) and seemed to be an organic and ambitious project from a political point of view. It took place in two phases, and, during the first, there were plans to provide the South with conditions and material infrastructures that would aid development. In the second phase, there would be intense public industrialization in strategic sectors (iron and steel, mechanics and chemical industries), mostly organized by the Ministry of State Participation, and in combination with a system of incentives and cooperation between local actors, entrepreneurs, and trade unions, working together to create local and peripheral agreements (Galasso 2009). The infrastructure policies were accomplished within the first few years of the Fund's existence (1950–1957); the active industrialization was realized in two different phases, between the years 1959–1965 and 1969–1973 (Giannola and Petraglia 2007). In the first phase, resources from the Marshall plan and from the World Bank for the South were used to make investments in basic infrastructure, such as transport routes, roads, ports, bridges, and also aqueducts, land reclamation, electrical networks, and schools. In the second phase, the process of industrialization began, based on the industrial development pole model (Perroux 1966), which chose to concentrate investments in certain areas of the South, favoring heavy industry, basic, and intermediate production, and which served as the raw materials for the already well-established industries in the Center-North. Between the years 1959 and 1965, there was greater consolidation of the basic sectors and intermediate production. This choice reflected a national political and economic strategy. In the division of work at a national level, the southern industries were seen as a support to those in the North (mostly the textile, mechanical, and car industries), which produced finished products to be introduced onto the market. They were therefore the protagonists during this intense phase of development and international affirmation.

The drawback to this type of industrialization process was that it focused on sectors that, once operational, would not need a large work force, would not produce a market-ready finished product, or trigger the creation of related industries. They also included specific processes that were wholly alien to the territories in which they were established, often completely erasing existing local production (as in the case of jasmine cultivation for the perfume industry in the area surrounding Milazzo, in the province of Messina, which was decimated by the construction of a petrochemical refinery). Not to mention the environmental effects of choosing to delocalize the most polluting industries to areas willing to make the trade-off between guaranteed work and land consumption, as Hytten and Marchioni (1970) observe in their classic book on the petrochemical industry in Gela.

In the South, the exchange between work and environmental impact not only involved heavy industry but also building speculation, which was facilitated by the pressures of the criminal and informal economy and by the local administrations' inability to impose clear laws and subsequently enforce them. Here, we are thinking of the unfortunate issue of development plans for southern cities (which were either left unapproved or duly ignored), of the building amnesties that made it easier to build illegally and "legitimize" subsequently, of the considerable consumption of agricultural land, especially close to urban centers and along the coasts, leading to the near disappearance of peri-urban farming. Over time, the building sector became the majority employer in many southern provinces, overtaking the industrial sector. Clientelistic bureaucracy and building were tied together in a particular growth model, based on the speculative consumption of two collective resources (territory and public investments), but this model was unsustainable in the long run and expensive in terms of how much it contributed to disfiguring the environment and affected the health of the population. These negative effects are still visible today in many southern areas, and Messina can be considered emblematic from this point of view (see Chap. 4).

These contradictions had already been condemned by Mottura and Pugliese who, in a study from 1975 (p. 190), reduced the extent of industrial growth created by the extraordinary intervention. Between 1951 and 1971, employment in the industrial sector in the South rose to just over 500,000 units, against the 1,500,000 in the Center-North. Half of this increase was attributable to the building sector. In other words, the older, more traditional sectors had been replaced by other sectors with low labor absorption (basic industries) or precarious and unstable work (building). In addition, there had also been excessive growth in the parasitic public tertiary sector, in order to guarantee consensus and social control. The main effect of the policies for the South was the "growth of a state bourgeoisie, which dispensed and was the beneficiary of public spending" (Mottura and Pugliese 1975, p. 190). This middle class (who were often former members of the agricultural class) played a particularistic role in the social redistribution of public finances, and they gained many advantages both in economic terms and in producing political consensus. Pizzorno (1980) shows how the Christian Democratic Party (Democrazia Cristiana or DC) played a very important role in consensus management after the Second World War. In order to counter the expansion of the Italian Communist Party (PCI) in the Center-North, the DC, especially in the South, pushed for the incorporation into the middle class of some of the more conservative social groups (particularly shopkeepers and public-sector employees). They did this through the particularistic and corporative inclusion of citizens, centered on the fragmented and categorical granting of rights (which often took the form of "favors") (Calise 1978; Barbagallo 1980; Trigilia 1992; Fantozzi 1993). Examples of this include the granting of building permits, the high tolerance of tax evasion in the retail trade, the introduction of advantageous pension schemes for social categories created ad hoc, and the definition of categories of special "workers" for whom jobs in the public administration were made available without the usual selection procedures. Such practices occurred within clientelistic-political networks where the broker was crucial for the success of the transaction, and the dual effect was the subordination of the co-opted person, and the affirmation of the lack of separation between the political and institutional spheres. The personal dependence of these networks on so-called mediators (Graziano 1974; Gribaudi 1980) had the twofold role of reinforcing the clientelistic mechanisms of local control and creating fragmentation and individualism, thus impeding the formation of a collective "class" consciousness (Pizzorno 1993).

However, because the clientelistic relationship is always dyadic, with the patron on one side and the potential clients on the other, the system works only if the patron is truly able to guarantee redistribution through the particularistic inclusion of the clients, even if this takes place at a future date. The greater the ability to expand particularistic inclusion to increasingly external social circles, the greater the social consensus will be, and the individual tendency to consider it both reasonable and convenient to become involved in such networks, albeit in a position of subordination. For this reason, political clientelism is a strategy that is more opportunistically pursued the more it is able to guarantee the possibility of social integration and mobility, regardless of whether or not this possibility is the result of an objective right or a particularistic appropriation. Once the action has been carried out, it tends to become an acquired right, and its "particularistic" and "selective" features become less obvious and increasingly difficult to demonstrate. In our view, the reason for the widespread practice of clientelism for social regulation, especially in the South, is to be found in its ability to extend its benefits to the more marginal or external social circles. In fact, when the possibility of integration is weakened due to austerity policies that curtail public spending (as was the case in Italy in the 1990s), clientelism, as a system for guaranteeing political consensus, appears increasingly flawed.

Moreover, the dual relationship between patron and client involves the active mobilization of the client, albeit from a subordinate position. The clients are therefore not simply passive users, but active participants in the mediation relationship (Boissevain 1966; Scott 1972; Signorelli 1983). Establishing a clientelistic network and continuing to be part of one implies certain relational costs that can only be sustained if they are adequately rewarded, in either material (e.g., by obtaining benefits) or immaterial (by increasing the powers of influence) terms. Thus, the ability to create a "connection" or to recognize the right connection is an essential strategy for the client.

The "right" connection is one that allows the client to do something that is directly linked to gaining a benefit, but it does not guarantee the outcome of the action/relationship, which therefore remains uncertain. This is especially true if we consider that clientelistic relationships are rarely direct, but always masked by mechanisms of instrumental friendship (Wolf 1966) that leave the terms of the exchange implicit or ambiguous. Getting on board the clientelistic "bandwagon" or—to use some local idioms—finding the right "road" to obtaining the desired benefits requires a huge effort and constant activity, which generally means finding the "right key" and making sure that it fits the "right door" and does not remain stuck in the wrong lock. The actors involved have to make use of the resources available to them, and in an environment such as the South, where these are limited, and where economic and cultural capital are scarce, their only choice is to use the social/relational resources open to them. In this way, the use of private relationships (family, relations, and friends) is a strategic resource for making effective decisions in the public sphere (Signorelli 1983).

3.5 Italian Neoliberalism

As seen in the previous paragraph, the extraordinary intervention almost immediately presented a series of contradictions and uncertainties. From the mid-1970s, these were accentuated by the economic choices Italy made in order to deal with the economic crisis (Cafiero 2000). Due to the high cost of crude oil, production in steel works and chemical plants became extremely expensive, and the mechanical industries, car, and plane manufacturing, for example, became obsolete or inefficient, as organizational and technological innovations led to a reduction in mass production, in favor of lean production. The introduction of automated machinery resulted in job loss and instead of supporting industrial re-qualification, the Italian State made a decision which had amplifying negative effects: that was the decision to interrupt offer-based policies (direct investments in the economy and infrastructure system), in favor of demand-based policies (indirect transfer of families and businesses). This was the moment when public spending in the South became a means for the unproductive support of southern incomes in exchange for electoral support (Giannola 2000; Giannola and Petraglia 2007).

The decision to abandon investment policies further weakened the fragile industrial base in the South, which was, in many cases, unable to diversify. Where it did manage to, it suffered the consequences of decline, in the form of downsizing or the miniaturization of businesses, the partial informalization of activities through recourse to undeclared work and, finally, scant regard for environmental and work laws.

The inability to create stable jobs within the local labor market became structurally entrenched and intensified social pressure, which was only partially eased by migration flows and informal work in the trade/sales sector, the private tertiary sector, and the building sector. The development of a particularistic welfare state, which sent an automatic flow of resources to the South, and the growing role of the public sector within the labor market (with public sector jobs filling the gaps in the private sector) acted as safety valves and helped to reduce some of the social tension.

The regression to a pathologically state-assisted economy lies at the basis of northern discontent, which, during the 1990s, led to the rise of the Northern League (Lega Nord), a group which predictably based its success on the stereotypical belief that the South is an unproductive cancer that inhibits the country's growth. As several authors have pointed out, this idea is ideological and untrue. In a single national market with a strongly dualistic structure, such as the Italian one, the artificial support of "demand" for the weaker, more backward side, does not work as a Keynesian multiplier, nor does it help create local "supply" to help develop production. On the contrary, it induces the stronger, more advanced, side to import, thus decreasing production and slowing down the formation of a marketplace for the weaker side (Giannola and Petraglia 2007, p. 21). The lack of a solid local production base associated with a rise in income, thanks to public spending, results in greater spending power for the citizens of the South, which in turn benefits the industries in the rest of the country (Bodo and Viesti 1997; Giannola 2000; Jossa 2001; Viesti 2003). Furthermore, until the beginning of the 1990s, public spending was financed by issuing money and an increase in the public debt (Viesti 2009), which, in an unstable international context caused by the end of the *Bretton Woods* agreement, gave rise to inflationary tendencies and repeated devaluations. In this way, a subtle balance is created and the situation becomes "convenient" not only for the assisted area, but also for the more industrialized one, which is stimulated to expand further, both internally (to cover the needs and demands of the southern market) and externally. In fact, the weak national currency supported exportation and contributed to the growth and stability of small northern businesses, which found themselves in a more dynamic context (a process that resembles, among other things, Gunder Frank's analysis, 1972). In turn, this stimulated changes in and the upgrading of the industrial fabric that were to be decisive for the future growth of the Italian economy, centered on the industrial districts and on the Italian brand of *Made in Italy*.

This is how the division of labor took place between the South, which was left to its "politics", and the North, which devoted itself to business (Trigilia 1992; Jossa 2001), determining a balance between the industrial interests of the Center-North which coincide with the political interests and consensus of the South.

The balance was broken during the transition to a post-Fordist neoliberal regime for social regulation, in which globalization processes (the growing interconnection between national economies and their financialization)[17] made public debt expansion policies impossible, and relegated the State from rules maker to rules taker.

On a national level, the scandals surrounding *Mani Pulite* (Clean Hands), and *Tangentopoli* (Bribesville), from 1992 to 1994, show a complex system of corruption involving political parties, public officials, and private enterprise. The corrupt made profits by operating in "protected" public markets (contract procurement, health care, state-owned enterprises), where bribes served to finance the party system but also to swell the personal accounts of politicians and administrators (among others see, Vannucci 2009, 2016).

Even though the Lega believed that the extraordinary intervention and the *Mezzogiorno* were the main reasons for the high level of corruption undermining the economic development of the country as a whole, the pervasive nature of the *Mani Pulite* scandal underlined the fact that corruption was a feature of the way the Keynesian-Fordist development model had taken root in all parts of the country. The *Tangentopoli* scandal affected all branches of the public sector on a national level. Most notable were the investigations concerning the health-care sector (e.g., the scandal over infected blood products); the Enimont Affair involving the ENI, Ente Nazionale Idrocarburi (National Hydrocarbons Agency), and the Ferruzzi Montedison group, the *Fiamme Sporche* investigation which led to the discovery of 80 corrupt officials within the financial police, and the investigation of Craxi's secret funds (with the discovery of a super-bribe of 10 billion lire paid by Berlusconi) (Travaglio et al. 2012).

The shock surrounding the Mani Pulite scandal accelerated and justified the processes of change already taking place within the economic and social fabric. Globally, at that time, a transition to a globalized post-Fordist regime was taking place, in which the financialization and de-materialization of the economy were becoming increasingly central, as were the neoliberal policies for the de-regulation of trade and labor, the flexibilization of production, and the increase in its market dependence (among others, Harvey 2010).

Neoliberal policies had arrived in Italy, supported by the imaginary of a serious political-economic crisis and the need to get over clientelism, welfarism, dependence, and increasing public debt and waste, in order to improve competitiveness and fiscal balance (among others, Harvey 2005). New public management measures (Klett 2005; Pollitt et al. 2007; Pollitt and Bouckaert 2017) were established to simplify public administration and make administrative activity more transparent, and introduce the principles of decentralization, autonomy and accountability, and control and monitoring of results, in order to render the system more efficient, effective, and economical. After the extraordinary intervention for the South, bottom-up development policies emerged strongly (Rossi 2004; Viesti 2011; Milio 2010, pp. 61–80), strengthening the intangible factors and material resources and improving its institutional abilities (Amin 1999; Trigilia 2001, 2005; Viesti 2003). However, these policies lacked a general macro-design and ended up becoming localistic (Viesti 2009; Trigilia 2012).

According to Rossi (2012), the two main pillars of Italian neoliberalization were the flexibilization of the labor force and the de-regulation of the housing market. The flexibilization of the labor force resulted in the introduction of numerous temporary contracts and the progressive informalization of work, but without providing social security provisions and benefits for these new professional figures. Flexibility within the labor market has increased precariousness without resolving the contradictions in the Italian labor market. Instead, it has become more polarized, between a limited circle of guaranteed workers (generally working in larger industries and the public sector) and a host of unprotected workers in the weaker sectors (private tertiary, sales, and small businesses) (Mingione 1991; Ferrera 1996; Molé 2010).

Moreover, as Rossi says (2012, pp. 5 e ss.):

> The second pillar of neoliberalism in Italy is the housing sector. The supply of public housing decreased by about 20% between 1991 and 2007. [...]. In the past, the historically limited supply of public and subsidized housing in Italy was compensated by legislation aimed at keeping rental prices affordable for disadvantaged groups in the private property market. [...] However, legislative acts passed by Parliament in 1992 and 1998 [...] gradually neutralized the 'fair rent' policy. [...]. As a result of this twofold process—the shrinkage of the public housing supply and the neutralization of regulations making the private rental sector more affordable—Italy has become a nation with a record number of homeowners: in 2010, two years after the end of the housing bubble of the 2000s, 84% of households were owner-occupiers. For many years, conventional wisdom has portrayed Italian families as tireless savers and far less as indebted homeowners [...]. [However], in late 2010 [...] a report by the Bank of Italy showed that the so-called "mortgage delinquency rate" had steadily increased among mortgaged Italian households between 2000 and 2007 (up to 5%), thus before the financial crisis, and in Europe, Italians are ranked at the top of the credit risk list along with Spaniards. The social groups that are most exposed to credit risk are those who have lost their jobs after stipulating a mortgage, but also single parents, low-income workers and the expanding category of temporary ('precarious') employees.

Over the past thirty years, precarious employment and inequalities in accessing housing, the reduction of the welfare state, and neoliberal policies have all had the effect of increasing precariousness, placing the weakest local and social groups in a position of social vulnerability (Ferrera and Gualmini 2004; Borghi and Rizza 2006; Paci 2006; Da Roit 2007; Ascoli 2011; Pau and Moreno-Fuentes 2013; Ascoli and Pavolini 2015) and reducing the Italian family's ability to compensate for social risks that is typical of the Mediterranean welfare model (Ferrera 1996; Saraceno 1994; Naldini 2003).

On one side, the constant national scandals reported in the papers linked to corruption and the mafia, and on the other, the neoliberal policies of austerity that have produced the general impoverishment of the country, especially among the weaker categories, have highlighted how the "Southern Question" theory within the North/South dichotomy is not sufficient to explain the Italian development process. Instead of being exceptions in the move toward modernity, the southern cities seem to be, in a certain sense, the precursors of contemporary capitalism.

3.6 Conclusion

In this chapter, the central concern is the *Mezzogiorno* and the "age-old question" (Gramsci 1966) of Italian history. In different ways and during diverse political and economic cycles, this can be seen in the constant disparity between the main economic indicators, in the backwardness of manufacturing facilities, and the inability to create stable employment, as well as in the protraction of informal employment, illegal building practices, clientelism, and opportunistic/clientelistic practices used to acquire public benefits. Through a strange sort of "orientalism" (Schneider 1998b), the Italian South is depicted as being impervious to modernization, with an innate attitude of inertia and even illegality. The southern "family", enclosed in its amoral familism, is seen as clan-like, its main purpose to safeguard the economic status and standing of its most immediate members. Over the course of the chapter, the aim has been to highlight how this analysis is both ideological and simplistic, and that the combination of clientelism and welfarism in the South is part of a development model that involves the whole of Italy.

In this sense, the southern territories and their history express fragments of modernity and tell us a great deal about the dynamics of contemporary capitalism. It is therefore possible to see how the building of the slums in Messina was a process that was produced and reproduced after the earthquake, as a means to manage the space economically and politically (see Chap. 4). The chapter continued by showing that in the agricultural question that was part of the Southern Question, the peasants were mixed figures who gradually became proletarianized laborers, and eventually a surplus labor force to be used as low-cost labor, as well as a source of electoral support. The chapter reviewed the processes that transformed the southern city from a medium- and wide-range commercial and industrial hub, connected to a vast agricultural system situated in the surrounding rural areas, to an exclusively bureaucratic center, disconnected from the production bases and dedicated to tertiary administrative sectors, with local commerce and building as the main industries. It also showed how the policies regarding the Fund for the South were ambivalent, in terms of planning and implementation.

As part of the Southern European welfare model (Ferrera 1996), the cycle of public spending had favored an increase in income levels, even though particularistic and clientelistic forms of inclusion had permitted redistribution. However, from the end of the 1990s, policies used to contain the deficit, such as fiscal decentralization and the liberalization of public services, broke this internal equilibrium based on public spending. The territory slowly began to feel the effects of post-Fordist capitalist policies (Harvey 1989a, 2005; Amin 1994; Jessop 2002; Mingione 2014). Beyond local development rhetoric, the territory was pushed further and further to the margins of the international work market as it became involved in global competition, mostly because it had few material or immaterial resources to put forward, and therefore used the only two it could offer—the land and the workforce—with the consequent long-term effects of impoverishment. This is well documented by the growing informalization of production, with the de-regulation of large areas of the work market and the commodification of more and more areas of material life (among others, Sassen 1997; Tabak and Crichlow 2000; Slavnic 2010; Breman and Linden 2014), as well as the increase in speculation linked to the territory (i.e., building speculation, rubbish disposal, the externalization of polluting processes, and land grabbing) (among others, see Araghi 2009; McMichael 2004; Moore 2015; Sevilla-Buitrago 2015).

Post-Fordist capitalism has shown its ability to create profit and to regenerate by inventing new commodification spaces, even creating new life experiences and the production of subjectivity through bio-capitalism (we are thinking here of *social networks*, or the rise of new themed *shopping malls*, for a *shopping experience*) (Ritzer 1993; Lazzarato 2012; Fumagalli 2015). The increase in profit margins derived from the constant commodification of new areas of life and the market can be associated with the ability of capitalism to incorporate some very different production regimes according to its need to accumulate capital. These can vary from classic pre- and proto-capitalistic forms, such as slave labor and piecework, to those branded as backward and underdeveloped, such as illegal work and underemployment, up to more modern self-disciplining techniques, such as the self-employment of flexible post-Fordist workers (Lazzarato 1996; Sennett 1998; Fisher 2009; Morini and Fumagalli 2010; Harvey 2010; Heinz 2010; Johnson 2016). This is increasingly evident in the complexification and de-standardization of work but also in the manipulation of the "territory" which, over the past few decades, has progressively become the object of capitalistic use (Moore 2016); the "production" of places for private use and consumption (e.g., phenomena such as urban sprawl, gentrification, or industrial delocalization in greenfield contexts, not forgetting heritagization processes) (among others, Harvey 1989b; Davis 1990, 2006b; Sorkin 1992; Zukin 2010; Bendix et al. 2017).

Braudel (1988) had already observed that the essence of capitalism is not the free market, but the ability to create profit by controlling long monopolistic networks, that necessarily move within very large spaces, where, from time to time, new portions are included in a subordinate way. In this respect, profit can be created, thanks to the social actor's centrality, ability to manage, control, and organize product flow and the supply chain by diversifying and dematerializing them (which has happened more recently with economic financialization). These long capitalistic chains are characterized by strongly asymmetrical power and unbalanced socio-economic relationships. Large profit margins also derive from the ability to intermediate. In short, in post-Fordist capitalism, the factors that had once been considered obsolete remnants from the past, and an expression of backwardness (e.g., different types of non-standard or informal work, personalistic intermediation, oligopolies that hinder free competition, etc.) were found to be essential for the creation of profit within increasingly segmented circuits. Along these long chains of global capitalism, a growing number of opportunities arise for creating an economy that is no longer based only on the production of material goods. Indeed, greater profits are to be made through an intermediation economy, where there are an increasing number of speculative movements, and value is created in the passage between one financial intermediary and another.

As is seen in the following chapters, the history of Messina is one of those spaces in which the strands of modernity come together, sometimes well ahead of the times.

Notes

1. Of great interest on this subject is the documentary made by Dickie and Sayer (2005), "Messina: a city without memory?"; see also works by Dickie (2000, 2002, 2008).
2. For Herzfeld (2005, p. 3), cultural intimacy is "the recognition of those aspects of a cultural identity that are considered a source of external embarrassment but that nevertheless provide insiders with their assurance of soci-

ality, the familiarity with the bases of power that may at one moment assure the disenfranchised a degree of creative irreverence and at the next moment reinforce the effectiveness of intimidation. Cultural intimacy may erupt into public life. This can take the form of ostentatious displays of those alleged national traits—American folksiness, British 'muddling through', Greek mercantile craftiness and sexual predation, or Israeli bluntness, to name just a few—that offer citizens a sense of defiant pride in the face of a more formal or official morality and, sometimes, of official disapproval as well. These are the self-stereotypes that insiders express ostensibly at their own collective expense. Cultural intimacy comes from the tension created between stereotypes and strategies of representation".

3. Following Giddens (1990, p. 92), an event such as an earthquake undermines that "ontological security" of subjects, defined as "feelings of security" [...] "refers to the confidence that most humans beings have in the continuity of their self-identity and in the constancy of the surrounding social and material environments of action. A sense of the reliability of persons and things, so central to the notion of trust, is basic to feelings of ontological security; hence the two are psychologically closely related. Ontological security has to do with 'being' or, in the terms of phenomenology, 'being-in-the-world'. But it is an emotional, rather than a cognitive, phenomenon, and it is rooted in the unconscious".
4. On Sicilianity and Sicilianism, see Schneider and Schneider (1994) and the observations made by Herzfeld (2005, p. 17). It is enough to remember here that the strength of stereotypes as essentialist, naturalist representations of identity is also that of being flexible, modifiable, and adaptable to the context, and according to the needs of the politically manipulable identity. Thus, the role of the researcher should be that of trying to dismantle "the game of reciprocal and dual operations for the construction of stereotypes and metaphors of belonging, showing the use of real political and discursive contexts" in order to give a critical reading of the processes of identity definition (Palumbo 2001, p. 126).
5. On shacks, suburbs, and Roma in recent Italian history, many of the observations expressed by Forgacs (2014) apply to the Messina case.
6. Benedetto Croce suggests that this expression dates back to the fourteenth century and was used for the first time by Tuscan merchants visiting the city of Naples (Croce 2009, p. 13).
7. These are the same stereotypical expressions used by the inhabitants of the city to describe the people living in the shantytowns in a Facebook page on the theme of rehousing, posted by the newly elected Mayor who, as soon as he took office, made several proposals for "reconstruction" and "rehousing", soliciting collective discussion on social networking sites. The Facebook page has received over 20,000 "Likes" and there are more than 21,000 followers, and it reaches many different social levels.

8. On this matter, see the pages in Polanyi on land and nature in *The Great Transformation* (1944) and Moore's more recent analysis (2015).
9. As highlighted by Escobar (1984, p. 381), continuing the Foucaultian analysis of power, discourse, and knowledge, "the history of Western and, increasingly, non-Western societies [...] is of a steady process of appropriation and consequent disposition of societal background practices, common social meanings and cultural contents through a series of discourses, institutions and practices—a process by which the material conditions of life and the unspoken mechanisms of culture are brought into the realm of explicit calculations and subjected to an infinite number of forms of power-knowledge".
10. The southern Italian regions of Campania, Basilicata, Calabria, Puglia Molise, Abruzzo, and the two islands—Sicily and Sardinia—have many internal differences, and it seems to be forcing the issue to analyze them as one single backward territory.
11. The southern intellectuals, far from having a subaltern or marginal role, contributed (and contribute) to debate and policies, often overturning, in a positive way, the ambivalent stereotype of the "southerner" and the South. They contrasted the idea of the bad savage with that of the good savage, showing an atavistic South that was both natural and bucolic, the depositary of an ancient wisdom and particular spirit that represented the South as slowness, idleness, creativity, solidarity, and family ties that become virtues in comparison to the frenzy of modern life (Demarco 2009; Teti 2011). This sort of analysis was, however, caught in the trap of ahistorical essentialization.
12. For an analysis of the dynamics behind the origins of international capitalism in Sicily, through supply chains that already existed during the fifteenth century, see, among others, Giarrizzo (1987, p. XLIX e ss.), Aymard (1987, pp. 5–40), and Morreale (2018); for Sicilian silk production: Laudani (1989); on the cultivation of citrus fruits Lupo (1984); for wheat, Aymard (1989); for sugar, Morreale (2006); and for sulphur, Blando (2009).
13. Aymard (1989, p. 770), for example, highlights the fact that wheat was a speculative monoculture linked to foreign markets, and following the modification of the international market (and unable to compete with the amount produced), the wheat sector tended to move to transforming the product into pasta, and to other "new agricultural specializations that were more profitable, such as citrus production and vineyards, although still driven by a speculative logic in their relation to national and international markets" (ibid., p. 779).
14. The debate surrounding the southern agrotown is considerable. Among others, see Blok (1969), and more recently Curtis (2013).

15. It is a very wide area of study; among others, see, Bagnasco (1977); Bagnasco and Pini (1981); Brusco (1982, 1986); Piore and Sabel (1984); Trigilia (1986); Blim (1990); Becattini (1991); Becattini et al. (2009).
16. About the Fund for the South, see among others, Cafiero (2000); OECD (2001, pp. 75–90); Martinelli (2009); Giannola and Petraglia (2007); Farinella (2010); Felice and Lepore (2016).
17. There is a huge amount of literature on the transition from Fordism to post-Fordism. Among these, see Harvey (1989a); Mingione (1991); Jessop (1992); Amin (1994); on the processes of globalization and financialization from a critical point of view, the work of Harvey (2010, Chapter 1); Fumagalli and Mezzadra (2010); and in particular the essays by Marazzi (2010); Lucarelli (2010); Paulré (2010); and Vercellone (2010).

References

Abu-Lughod, J. (1989). *Before European Hegemony: The World System A.D. 1250–1350*. New York: Oxford University Press.

Agnoletti, M. (Ed.). (2013). *Italian Historical Rural Landscapes*. London and New York: Springer Dordrecth Heidelberg.

Agrawal, A. (1995). Dismantling the Divide Between Indigenous and Scientific Knowledge. *Development and Change, 26*, 413–439.

Amin, A. (Ed.). (1994). *Post-fordism. A Reader*. Oxford: Blackwell Publishers.

Amin, A. (1999). An Institutionalist Perspective on Regional Economic Development. *International Journal of Urban and Regional Research, 23*(2), 365–378.

Anderson, B. (1991). *Imagined Communities: Reflections on the Origin and Spread of Nationalism* (Revised and extended ed.). London: Verso.

Antweiler, C. (1993). Local Knowledge and Local Knowing: An Anthropological Analysis of Contested 'Cultural Products' in the Context of Development. *Anthropos, 93*, 469–494.

Araghi, F. (2009). Accumulation by Displacement. Global Enclosures, Food Crisis and the Ecological Contradiction of Capitalismo. *Review (Fernand Braudel Center), 32*(1), 113–146.

Arrighi, G., & Piselli, F. (1987). Capitalist Development in Hostile Environments: Feuds, Class Struggles, and Migrations in a Peripheral Region of Southern Italy. *Review (Fernand Braudel Center), 10*(4), 649–751.

Ascoli, U. (Ed.). (2011). *Il welfare in Italia*. Bologna: Il Mulino.

Ascoli, U., & Pavolini, E. (Eds.). (2015). *The Italian Welfare State in a European Perspective. A Comparative Analysis*. Bristol: Policy Press.

Aymard, M. (1987). Economia e società, uno sguardo d'insieme. In M. Aymard & G. Giarrizzo (Eds.), *La Sicilia* (pp. 5–40). Torino: Einaudi.

Aymard, M. (1989). Il sud e i circuiti del grano. In P. Bevilacqua (Ed.), *Storia dell'agricoltura itaiana, vol. 3* (pp. 755–787). Venezia: Marsilio Editore.
Bagnasco, A. (1977). *Tre Italie: la problematica territoriale dello sviluppo italiano.* Bologna: Il Mulino.
Bagnasco, A., & Pini, R. (1981). *Sviluppo economico e trasformazioni sociopolitiche dei sistemi territoriali a economia diffusa* (p. 14). Milano: Quaderni Fondazione Giangiacomo Feltrinelli.
Bagnasco, A., Messori, M., & Trigilia, C. (1978). *Le problematiche dello sviluppo italiano.* Milano: Feltrinelli.
Banfield, E. C. (1958). *The Moral Basis of a Backward Society.* Glencoe: The Free Press.
Barbagallo, F. (1973). *Lavoro e esodo nel Sud 1861–1971.* Napoli: Guida.
Barbagallo, F. (1980). *Mezzogiorno e questione meridionale (1860–1980).* Napoli: Guida.
Barbagallo, F. (1994). *La modernità squilibrata del Mezzogiorno d'Italia.* Torino: Einaudi.
Barone, G. (1989). Mezzogiorno ed Egemonie Urbane. *Meridiana, 5*, 13–47.
Becattini, G. (1991). Italian Industrial Districts: Problems and Perspectives. *International Studies of Management & Organization, 21*(1), 83–90.
Becattini, G., Bellandi, M., & De Propris, L. (2009). *A Handbook of Industrial Districts.* Cheltenham: Edward Elgar.
Bendix, R. F., Eggert, A., & Peselmann, A. (2017). *Heritage Regimes and the State.* Gottingen: Gottingen University Press.
Bevilacqua, P. (1984). Quadri mentali, cultura e rapporti simbolici della società rurale del Mezzogiorno. *Italia Contemporanea, 154*, 51–70.
Bevilacqua, P. (1989). *Storia dell'agricoltura italiana in età contemporanea, vol. III.* Venezia: Marsilio Editore.
Bevilacqua, P. (1993). *Breve storia dell'Italia Meridionale dall'Ottocento ad oggi.* Roma: Donzelli.
Bhabha, H. (1990). The Other Question: Difference, Discrimination, and the Discourse of Colonialism. In R. Ferguson, M. Gever, T. T. Minh-ha, & C. West (Eds.), *Out There: Marginalization and Contemporary Cultures* (pp. 71–89). New York; Cambridge: New Museum of Contemporary Art; MIT Press.
Blando, A. (2009). Da un monopolio naturale all'altro: il grano e lo zolfo siciliano. In S. Biagio (Ed.), *Lo spazio tirrenico nella "grande trasformazione". Merci, uomini e istituzioni nel settecento e nel primo ottocento* (pp. 1–39). Bari: Edipuglia.
Blim, M. (1990). *Made in Italy: Small-Scale Industrialization and Its Consequences.* New York: Praeger Publishers.
Blok, A. (1969). South Italian Agro-Towns. *Comparative Studies in Society and History, 11*(2), 121–135.
Blok, A. (1974). *The Mafia of a Sicilian Village. 1860–1960.* New York: Haprer & Row Publishers.

Bodo, G., & Viesti, G. (1997). *La grande svolta. Il Mezzogiorno nell'Italia degli anni novanta*. Roma: Donzelli.
Boissevain, J. (1966). Patronage in Sicily. *Man, 1*(1), 18–33.
Borghi, V., & Rizza, R. (2006). *L'organizzazione sociale del lavoro. Lo statuto del lavoro e le sue trasformazioni*. Milano: Bruno Mondadori.
Braudel, F. (1982, V ed., ed.or.1949). *La Méditerranée et le monde méditerranéen à l'époque de Philippe II*. Paris: Armand Colin.
Braudel, F. (1985). *La dynamique du capitalisme*. Paris: Arthaud.
Braudel, F. (1997). Histoire, mesure du monde. In T. de Ayala & P. Braudel (Eds.), *Les Ambtions de l'Histoire*. Paris: Editions de Fallois.
Breman, J., & van der Linden, M. (2014). Informalizing the Economy: The Return of the Social Question at a Global Level. *Development and Change, Forum, 45*(5), 920–940.
Brusco, S. (1982). The Emilian Model: Productive Decentralisation and Social Integration. *Cambridge Journal of Economics, 6*, 167–184.
Brusco, S. (1986). Small Firms and Industrial Districts: The Experience of Italy. In D. Keeble & E. Wever (Eds.), *New Firms and Regional Development in Europe* (pp. 184–201). London: Croom Helm.
Cafagna, L. (1988). Modernizzazione attiva e modernizzazione passiva. *Meridiana, 2*, 229–240.
Cafiero, S. (2000). *Storia dell'intervento straordinario nel Mezzogiorno (1950–1993)*. Manduria, Bari, and Roma: Lacaita Editore.
Calise, M. (1978). *Il sistema DC. Mediazione e conflitto nelle campagne democristiane*. Bari: De Donato.
Caminiti, L. (2009). *La grande diaspora. 28 dicembre 1908. La politica dei soccorsi tra carità e bilanci*. Messina: GBM.
Croce, B. (2009). *Un paradiso abitato da diavoli*. Milano: Adelphi.
Curtis, D. (2013). Is There an 'Agro-Town' Model for Southern Italy? Exploring the Diverse Roots and Development of the Agro-Town Structure Through a Comparative Case Study in Apulia. *Continuity and Change, 28*(3), 377–419.
Da Roit, B. (2007). Changing Intergenerational Solidarities Within Families in a Mediterranean Welfare State. Elderly Care in Italy. *Current Sociology, 55*(2), 251–269.
Daniele, V., & Malanima, P. (2007). Il prodotto delle regioni e il divario Nord-Sud in Italia (1861–2004). *Rivista di Politica Economica, 3–4*, 267–315.
Daniele, V., & Malanima, P. (2014). Falling Disparities and Persisting Dualism: Regional Development and Industrialisation in Italy, 1891–2001. *Investigaciones de Historia Económica Economic History Research, 10*, 165–176.
Davis, M. (1990). *City of Quartz, Excaving the Future in Los Angeles*. London and New York: Verso.
Davis, J. A. (1998). Casting Off the 'Southern Problem': Or the Peculiarities of the South Reconsidered. In J. Schneider (Ed.), *Italy's "Southern Question". Orientalism in One Country* (pp. 205–224). Oxford: Berg.

Davis, M. (2006a). Bridging the Gap or Crossing a Bridge? Indigenous Knowledge and the Language of Law and Policy. In F. Berkes, D. Capistrano, W. V. Reid, & T. Wilbanks (Eds.), *Bridging Scales and Knowledge Systems: Concepts and Applications in Ecosystem Assessments* (pp. 145–182). Washington, DC: Island Press.

Davis, M. (2006b). *Planet of Slums*. London and New York: Verso.

De Martino, E. (2002). *La fine del mondo*. Torino: Einaudi.

Dei, G., Sefa, J., Hall, B. L., & Goldin Rosenberg, D. (Eds.). (2000). *Indigenous Knowledges in Global Contexts: Multiple Readings of Our World*. Toronto: University of Toronto Press.

Demarco, M. (2009). *Bassa Italia. L'antimeridionalismo della sinistra meridionale*. Napoli: Guida.

Dickie, J. (1999). *Darkest Italy. The Nation and Stereotypes of the Mezzogiorno, 1860–1900*. New York: St. Martin's Press.

Dickie, J. (2000). A Patriotic Disaster: The Messina-Reggio Calabria Earthquake of 1908. In G. Bedani & B. Haddock (Eds.), *The Politics of Italian National Identity. A Multidisciplinary Perspective* (pp. 50–71). Swansea: University of Wales Press.

Dickie, J. (2002). The Smell of Disaster: Scenes of Social Collapse in the Aftermath of the Messina – Reggio Calabria Earthquake, 1908. In J. Dickie, J. Foot, & F. Snowden (Eds.), *Disastro! Disasters in Italy Since 1860. Culture, Politics, Society* (pp. 235–255). New York: Palgrave.

Dickie, J. (2008). *Una catastrofe patriottica 1908: il terremoto di Messina*. Roma-Bari: Laterza.

Dickie, J., & Sayer, D. (2005). Messina: A City Without Memory?, Documentary. Retrieved from https://www.youtube.com/watch?v=vNU26Z9J1ms, http://www.ucl.ac.uk/place-and-memory/messina/index.html.

Escobar, A. (1984). Discourse and Power in Development; Michel Foucault and the Relevance of His Work to the Third World. *Alternatives, 10*(3), 377–400.

Escobar, A. (1995). *Encountering Development: The Making and Unmaking of the Third World*. Princeton: Princenton University Press.

Esteva, G. (1992). Development. In W. Sachs (Ed.), *The Development Dictionary. A Guide to Knowledge as Power* (pp. 6–25). London: Zeed Books.

Faeta, F. (2009). A Sud di nessun Nord. Per una critica delle usuali dislocazioni della geografia simbolica. In M. Petrusewicz, J. Schneider, & P. Schneider (Eds.), *I Sud. Conoscere, capire, cambiare* (pp. 147–162). Bologna: il Mulino.

Fantozzi, P. (1993). *Politica, clientela e regolazione sociale, il Mezzogiorno nella questione politica italiana*. Soveria Mannelli: Rubettino.

Farinella, D. (2010). *Mezzogiorno alla finestra. Società, territori e politiche nel Sud*. Roma: Aracne.

Farinella, D., & Ofria, F. (2016). Convergence in Italy Between the South and the Centre-North: After the Euro (Years 2002–2013). In G. Centorrino (Ed.), *Economia e Territorio, Profili economici, aziendali e statistici* (pp. 303–323). Roma: Aracne.

Fassin, D. (2011). *Humanitarian Reason. A Moral History of the Present*. Berkeley and Los Angeles: University of California Press.
Felice, E. (2013). *Perché il Sud è rimasto indietro*. Bologna: il Mulino.
Felice, E., & Lepore, A. (2016). State Intervention and Economic Growth in Southern Italy: The Rise and Fall of the Cassa per il Mezzogiorno (1950–1986). *MPRA Paper No. 69466*, 1–60.
Felice, E., & Vasta, M. (2015). Passive Modernization? The New Human Development Index and Its Components in Italy's Regions (1871–2007). *European Review of Economic History, 19*(1), 44–66.
Felice, E., & Vecchi, G. (2015). Italy's Growth and Decline, 1891–2011. *Journal of Interdisciplinary History, 14*, 507–548.
Ferrera, M. (1996). The 'Southern Model' of Welfare in Social Europe. *Journal of European Social Policy, 6*(1), 17–37.
Ferrera, M., & Gualmini, E. (2004). *Rescued by Europe?, Social and Labour Market Reforms in Italy from the First Republic to Berlusconi*. Amsterdam: Amsterdam University Press.
Fisher, M. (2009). *Capitalist Realism: Is There No Alternative?* Winchester: Zero Books.
Forgacs, D. (2014). *Italy's Margins. Social Exclusion and National Formation Since 1861*. Cambridge, UK: Cambridge University Press.
Foucault, M. (1997). *1975–1976: "Il faut défendre la société"*. Paris: Gallimard.
Fumagalli, A. (2015). The Concept of the Subsumption of Labour under Capital: Life Subsumption in Cognitive-Biocapitalism. In E. Fisher & C. Fuchs (Eds.), *Reconsidering Value and Labour in the Digital Age* (pp. 224–245). London: Palgrave Macmillan.
Fumagalli, A., & Mezzadra, S. (Eds.). (2010). *Crisis in the Global Economy*. Los Angeles, CA: Semiotexte.
Galasso, G. (2009). Fare un po' il punto sul Mezzogiorno. In E. Giammattei & P. De Vivo (Eds.), *Federalismo e Mezzogiorno*. Guida: Napoli.
Germani, G. (1973). Introduction. In G. Germani (Ed.), *Modernization, Urbanization, & the Urban Crisis* (pp. 3–57). Boston: Little, Brown & C.
Giannola, A. (2000). La nuova programmazione. Evoluzione e restaurazione. *Rivista Economica del Mezzogiorno, 3*, 403–422.
Giannola, A., & Petraglia, C. (2007). Politiche dell'offerta, della domanda e programmazione dello sviluppo. Il dualismo "dimenticato". *Rivista economica del Mezzogiorno, 1*, 13–41.
Giarrizzo, G. (1987). Introduzione. In M. Aymard & G. Giarrizzo (Eds.), *Storia d'Italia. Le regioni dall'unità ad oggi: la Sicilia* (pp. XIX–LVII). Torino: Einaudi.
Giarrizzo, G. (1992). *Mezzogiorno senza meridionalismo: la Sicilia, lo sviluppo, il potere*. Venezia: Marsilio Editore.
Giarrizzo, G. (2010). Conclusioni. In L. Caminiti (Ed.), *"Il disastro è immenso e molto più grande di quanto si possa immaginare". Il sisma calabro-siculo del 1908* (pp. 211–224). Aracne: Roma.

Giarrizzo, G., & Maraini, F. (1980). *Civiltà contadini ed immagini del Mezzogiorno negli anni '50*. Bari: De Donato editore.
Giddens, A. (1990). *Modernity and Self Identity: Self and Society in the Late Modern Age*. Stanford: Stanford University Press.
Gramsci, A. (1966). *La questione meridionale*. Roma: Editori Riuniti.
Gramsci, A. (1975). *Quaderni dal carcere. Vol.3, quaderno 19 (X) 1934–1935. Risorgimento italiano* (pp. 1959–2078). Torino: Einaudi.
Graziano, L. (Ed.). (1974). *Clientelismo e mutamento politico*. Milano: FrancoAngeli.
Gribaudi, G. (1980). *Mediatori. Antropologia del potere democristiano*. Torino: Rosenberg & Sellier.
Gribaudi, G. (1997). Images of the South: The Mezzogiorno as Seen by Insiders and Outsiders. In R. Lumley & J. Morris (Eds.), *The New History of the Italian South. The Mezzogiorno Revisited* (pp. 83–113). Exeter: Exeter University Press.
Gribaudi, G. (2010). Terremoti. Esperienza e memoria. *Parole Chiave, 44*(2), 85–105.
Gruzinski, S. (1988). *La colonisation de l'imaginaire. Sociétés indigènes et occidentalisation dans le Mexique espagnol XVIe–XVIIIe siècle*. Paris: Editions Gallimard.
Gunder Frank, A. (1972). *The Development of Underdevelopment*. Boston: New England Free Press.
Hardin, G. (1968). The Tragedy of the Commons. *Science, 162*(3859), 1243–1248.
Harvey, D. (1989a). *The Condition of Postmodernity. An Enquiry into the Origins of Cultural Change*. Cambridge: Blackwell Publishers.
Harvey, D. (1989b). *The Urban Experience*. Baltimore: Johns Hopkins University Press.
Harvey, D. (2005). *Brief History of Neoliberalism*. Oxford: Oxford University Press.
Harvey, D. (2010). *The Enigma of Capital and the Crises of Capitalism*. Oxford: Oxford University Press.
Heinz, R. K. (2010). Global Crisis – Global Proletarianization – Counterperspectives. In A. Fumagalli & S. Mezzadra (Eds.), *Crisis in the Global Economy* (pp. 197–236). Los Angeles, CA: Semiotexte.
Herzfeld, M. (2005). *Cultural Intimacy*. New York: Routledge.
Herzfeld, M. (2009). Le geografie simboliche dell'ineguaglianza: Roma capitale tra gli stereotipi di settentrionalità istituzionale e meridionalità sociale. In M. Petrusewicz, J. Schneider, & P. Schneider (Eds.), *I Sud. Conoscere, capire, cambiare* (pp. 131–146). Il Mulino: Bologna.
Hilowitz, J. (1976). *Economic Development and Social Change in Sicily*. Cambridge: Schenkman Publishing Co.
Hytten, E., & Marchioni, M. (1970). *Industrializzazione senza sviluppo. Gela: una storia meridionale*. Milano: FrancoAngeli.

Inea. (1976). *La Carta dello Sviluppo Agricolo (anni 1960–1969)*. Roma: Inea.

Jessop, B. (1992). Fordism and Post-Fordism: A Critical Reformulation. In A. J. Scott & M. J. Storper (Eds.), *Pathways to Regionalism and Industrial Development* (pp. 43–65). London: Routledge.

Jessop, B. (2002). *The Future of the Capitalist State*. Cambridge: Polity Press.

Johnson, M. (Ed.). (2016). *Precariat: Labour, Work and Politics*. London: Routledge.

Jossa, B. (2001). Il Mezzogiorno e lo sviluppo dall'alto. *Rivista economica del Mezzogiorno, 3*, 399–430.

Klein, N. (2007). *The Shock Doctrine. The Rise of Disaster Capitalism*. New York: Henry Holt and Company.

Klett, D. F. (2005). *The Global Public Management Revolution* (2nd ed.). Washington, DC: Brookings Institution Press.

Laudani, S. (1989). "Li posti delli mangani". Note sulla seta siciliana tra Sette e Ottocento. *Meridiana, 7*(6), 109–144.

Lazzarato, M. (1996). Immaterial Labor. In P. Colilli, E. Emory, M. Boscagli, C. Casarino, M. Turits, P. Virno, & M. Hardt (Eds.), *Radical Thought in Italy: A Potential Politics* (pp. 133–148). Minnesota: University of Minnesota Press.

Lazzarato, M. (2012). *The Making of the Indebted Man. Essay on the Neoliberal Condition*. Cambridge, MA: Semiotexte/MIT Press.

Lewis, O. (1959). *Five Families. Mexican Case Studies in the Culture of Poverty*. New York: Basic Books.

Leys, C. (2005). The Rise and Fall of Development Theory. In M. Edelman & A. Haugerud (Eds.), *The Anthropology of Development and Globalization* (pp. 109–125). Oxford: Blackwell.

Li Causi, L. (1976). Antropologia e ideologia. Note sul "patronage" nelle società mediterranee. *Rassegna Italiana di Sociologia, 18*(*1*), 119–131.

Lorenzoni, G. (1910). Sicilia, vol. 6. In G. Bertero (Ed.), *Italia. Giunta parlamentare d'inchiesta sulle condizioni dei contadini nelle province meridionali e nella Sicilia*. Roma: Tipografia nazionale.

Lucarelli, S. (2010). Financialization as Biopower. In A. Fumagalli & S. Mezzadra (Eds.), *Crisis in the Global Economy* (pp. 119–138). Los Angeles, CA: Semiotexte.

Lumley, R., & Morris, J. (Eds.). (1997). *New History of Italian South, The Mezzogiorno Revisited. New History of Italian South*. Exeter and Devon: University of Exeter Press.

Lupo, S. (1984). Agricoltura ricca nel sottosviluppo. Storia e mito della Sicilia agrumaria (1860–1950). *Archivio storico per la Sicilia orientale, 1*, 7–158.

Lupo, S. (1987). Tra società locale e commercio a lunga distanza: la vicenda degli agrumi siciliani. *Meridiana, 1*, 81–112.

Lupo, S. (1998). Storia del Mezzogiorno, questione meridionale, meridionalismo. *Meridiana, 32*, 17–52.

Lupo, S. (2015). *La questione. Come liberare la storia del Mezzogiorno dagli stereotipi*. Roma: Donzelli.

Malanima, P. (2003). Measuring the Italian Economy 1300–1861. *Rivista di Storia Economica, III*, 265–295.

Marazzi, C. (2010). The Violence of Financial Capitalism. In A. Fumagalli & S. Mezzadra (Eds.), *Crisis in the Global Economy* (pp. 17–60). Los Angeles, CA: Semiotexte.

Martinelli, F. (2009). Cassa per il Mezzogiorno. In R. Kitchin & N. J. Thrift (Eds.), *International Encyclopedia of Human Geography* (pp. 446–455). Amsterdam: Elsevier.

McMichael, P. (2004 third ed., 1996). *Development and Social Change. A Global Perspective*. Thousand Oaks: Pine Forge Press.

Meloni, B. (1984). *Famiglie di pastori: continuità e mutamenti in una comunità della Sardegna Centrale 1950–1970*. Torino: Rosemberg & Sellier.

Meloni, B. (1997). Famiglia meridionale senza familismo. Introduzione. In B. Meloni (Ed.), *Famiglia meridionale senza familismo* (pp. V–LXIII). Roma: Donzelli.

Milio, S. (2010). *From Policy to Implementation in the European Union: The Challenge of a Multi-Level Governance System*. London: I.B.Tauris.

Mingione, E. (1991). *Fragmented Societies: A Sociology of Economic Life Beyond the Market Paradigm*. Oxford: Blackwell.

Mingione, E. (2014). Le dinamiche del capitalismo, la mercificazione globale e la crisi economica: quali prospettive per il welfare delle società industriali avanzate. *Politiche Sociali, 2*, 201–215.

Moe, N. (2002). *The View from Vesuvius: Italian Culture and the Southern Question*. Berkeley, CA: University of California Press.

Molé, N. (2010). Precarious Subjects: Anticipating Neoliberalism in Northern Italy's Workplace. *America Anthropologist, 112*(1), 38–53.

Moore, J. W. (2015). *Capitalism in the Web of Life: Ecology and the Accumulation of Capital*. London: Verso.

Moore, J. W. (Ed.). (2016). *Anthropocene or Capitalocene? Nature, History, and the Crisis of Capitalism*. Oakland, CA: PM Press.

Morini, C., & Fumagalli, A. (2010). Life Put to Work: Towards a Life Theory of Value. *Ephemera, Theory & Politics in Organization, 10*(3/4), 234–252.

Morreale, A. (2006). *Insula Dulcis. L'industria della canna da zucchero in Sicilia, dal XV al XVII secolo*. Napoli: Edizioni scientifiche italiane.

Morreale, A. (2018). Capitalismo in Sicilia. In *Grano, zucchero e seta nei secoli XV–XVII*. Soveria Mannelli: Rubbettino.

Morris, J. (1997). Challenging "Meridionalismo" – Constructing a New History for Southern Italy. In R. Lumley & J. Morris (Eds.), *New History Of Italian South, The Mezzogiorno Revisited* (pp. 1–19). Exeter and Devon: University of Exeter Press.

Mottura, G., & Pugliese, E. (1975). *Agricoltura, Mezzogiorno e mercato del lavoro*. Bologna: Il Mulino.

Mutti, A. (1994). Il particolarismo come risorsa. Politica ed economia nello sviluppo abruzzese. *Rassegna Italiana di Sociologia, 4*, 451–518.

Naldini, M. (2003). *The Family in the Mediterranean Welfare States*. London and Portland: Frank Cass.

Noto, A. G. (2008). *Messina 1908. I disastri e la percezione del terrore nell'evento terremoto*. Soveria Mannelli: Rubbettino.

OECD. (2001). *OECD Territorial Reviews: Italy*. Paris: OECD Publication Service.

Paci, M. (2006). *Nuovi lavori, nuovo Welfare. Sicurezza e libertà nella società attiva*. Bologna: Il Mulino.

Palumbo, B. (2001). Campo intellettuale, potere e identità tra contesti locali, "pensiero meridiano" e "identità meridionale". *La ricerca Folklorica, 43*, 117–134.

Pantaleoni, M. (1891). Delle Regioni d'Italia in ordine alla loro ricchezza ed al loro carico tributario. *Giornale degli economisti, 2*(2), 48–88.

Pau, M., & Moreno-Fuentes, F. J. (2013). The Southern European Welfare Model in the Post-industrial Order. *European Societies, 15*(4), 475–492.

Paulré, B. (2010). Cognitive Capitalism and the Financialization of Economic Systems. In A. Fumagalli & S. Mezzadra (Eds.), *Crisis in the Global Economy* (pp. 171–196). Los Angeles, CA: Semiotexte.

Perroux, F. (1966). *L'economia del XX secolo*. Milano: Edizioni di Comunità.

Petrusewicz, M. (1996). *Latifondium. Moral Economy and Material life in a European Periphery*. Ann Arbor: University of Michigan Press.

Petrusewicz, M. (1998a). *Come il Meridione divenne una questione. Rappresentazioni del Sud prima e dopo il quarantotto*. Soveria Mannelli: Rubbettino.

Petrusewicz, M. (1998b). Before the Southern Question: "Native" Ideas on Backwardness and Remedies in the Kingdom of Two Sicilies, 1815–1849. In J. Schneider (Ed.), *Italy's Southern Question. Orientalism in one Country* (pp. 27–49). Oxford and New York: Berg.

Pezzino, P. (1993). *Il paradiso abitato dai diavoli: società, élites, istituzioni nel Mezzogiorno contemporaneo*. Milano: FrancoAngeli.

Piattoni, S. (1998). "Virtuous Clientelism": The Southern Question Resolved? In J. Schneider (Ed.), *Italy's "Southern Question". Orientalism in one Country* (pp. 225–243). Oxford and New York: Berg.

Piattoni, S. (2007). *Le virtù del clientelismo*. Bari and Roma: Laterza.

Pines, J. (2012). *The Art of Making Do in Naples*. Minnesota: University of Minnesota Press.

Piore, M., & Sabel, C. (1984). *The Second Industrial Divide: Possibilities for Prosperity*. New York: Basic Books.

Pizzorno, A. (1980). *I soggetti del pluralismo*. Bologna: Il Mulino.

Pizzorno, A. (1993). Identità e scambio politico nel conflitto industriale. In A. Pizzorno (Ed.), *Le radici della politica assoluta* (pp. 204–231). Milano: Feltrinelli.

Placanica, A. (1998). L'identità meridionale. *Meridiana, 32*, 153–182.

Polanyi, K. (1944). *The Great Transformation*. New York: Farrar & Rinehart.

Pollitt, C., & Bouckaert, G. (2017). *Public Management Reform. A Comparative Analysis – Into the Age of Austerity* (4th ed.). Oxford: Oxford University Press.

Pollitt, C., Thiel, S. V., Homburg, V., & Van Thiel, S. (Eds.). (2007). *New Public Management in Europe. Adaptation and Alternatives*. Palgrave Macmillan: New York.

Putnam, R. D. (1993). *Making Democracy Work: Civic Traditions in Modern Italy*. Princeton: Princeton University Press.

Ritzer, G. (1993). *The McDonaldization of Society: An Investigation into the Changing Character of Contemporary Social Life*. Newbury Park, CA: Fine Forge Press.

Rossi, U. (2004). New Regionalism Contested: Some Remarks in Light of the Case of the Mezzogiorno of Italy. *International Journal of Urban and Regional Research, 28*(2), 466–476.

Rossi, U. (2012). There's No Hope: The Global Economic Crisis and the Politics of Resistance in Southern Europe. *Belgeo, Revue belge de géographie, 1–2*, 1–18.

Rossi-Doria, M. (1948). Struttura e problemi dell'agricoltura meridionale. In M. Rossi-Doria (Ed.), *Riforma agraria e azione meridionalistica*. Bologna: Edizioni Agricole.

Rossi-Doria, M. (Ed.). (1969). *L'analisi zonale dell'agricoltura italiana*. Roma: Inea.

Salvemini, B. (1989). Prima della Puglia. Terra di Bari e il sistema regionale in età moderna. In L. Masella & B. Salvemini (Eds.), *Storia d'Italia. La Puglia* (pp. 1–218). Torino: Einaudi.

Saraceno, C. (1994). The ambivalent familism of the Italian Welfare State. *Social Politics, 1*, 60–82.

Saraceno, P. (2005). *Il nuovo meridionalismo*. Quaderni del Trentennale, 6. Napoli: Istituto Italiano per gli Studi Filosofici.

Sassen, S. (1997). *Informalization in Advanced Market Economies*. Issues in Development, Discussion Paper n. 20. Geneva: International Labour Office, Development Policies Department.

Schneider, J. (1998a). Introduction: The Dynamics of Neo-orientalism in Italy (1848–1995). In J. Schneider (Ed.), *Italy's "Southern Question": Orientalism in One Country* (pp. 1–23). Oxford and New York: Berg.

Schneider, J. (Ed.). (1998b). *Italy's "Southern Question": Orientalism in One Country*. Oxford and New York: Berg.

Schneider, J., & Schneider, P. (1976). *Culture and Political Economy in Western Sicily*. New York: Academic Press.

Schneider, J., & Schneider, P. (1994). Mafia, Antimafia, and the Question of Sicilian Culture. *Politics and Society, 22*, 237–258.

Scott, J. (1972). Patron-Client Politics and Political Change in Southeast Asia. *The American Political Science Review, 66*(1), 91–113.

Sennett, R. (1998). *The Corrosion of Character: The Personal Consequences of Work in the New Capitalism.* London and New York: W. W. Norton & Company.
Sereni, E. (1972). *Storia del paesaggio agrario.* Bari and Roma: Laterza.
Sereni, E. (1974). *Capitalismo e mercato nazionale.* Roma: Editori Riuniti.
Sereni, E. (1977, ed or. 1947). *Il capitalismo nelle campagne, 1860–1900.* Torino: Einaudi.
Sevilla-Buitrago, A. (2015). Capitalist Formations of Enclosure: Space and the Extinction of the Commons. *Antipode, 47*(4), 999–1020.
Shiva, V. (1989). *Staying alive: Women, Ecology, and Development.* London: Zed Books.
Signorelli, A. (1983). *Chi può e chi aspetta. Giovani e clientelismo in un'area interna del Mezzogiorno.* Napoli: Liguori.
Silverman, S. F. (1974). Patroni tradizionali come mediatori fra comunità e nazione: il caso dell'Italia centrale (1860–1945). In L. Graziani (Ed.), *Clientelismo e mutamento politico* (pp. 204–221). Milano: FrancoAngeli.
Slavnic, Z. (2010). Political Economy of Informalization. *European Societies, 12*(1), 3–23.
Sorkin, M. (Ed.). (1992). *Variations on a Theme Park: The New American City and the End of Public Space.* New York: Hill and Wang (Farrar, Straus and Girou).
Stiglitz, J, Sen. A., Fitoussi, J. P. (2009). *International Commission on Measurement of Economic Performance and Social Progress.* Retrieved from http://www.stiglitz-sen-fitoussi.fr/.
Tabak, F., & Crichlow, M. A. (2000). *Informalization: Process and Structure.* Baltimore: Johns Hopkins University Press.
Tarrow, S. (1967). *Peasant Communism in Southern Italy.* New Haven and London: Yale University Press.
Teti, V. (2011). *La razza maledetta. Origini del pregiudizio antimeridionale.* Roma: Manifestolibri.
Teti, V. (2013). *Maledetto Sud.* Torino: Einaudi.
Travaglio, M., Barbacetto, G., & Gomez, P. (2012). *Mani pulite. La vera storia.* Milano: Chiarelettere.
Trigilia, C. (1986). Small-Firm Development and Political Subcultures in Italy. *European Sociological Review, 2*(3), 161–175.
Trigilia, C. (1992). *Sviluppo senza autonomia. Effetti perversi delle politiche del mezzogiorno.* Bologna: Il Mulino.
Trigilia, C. (2001). Social Capital and Local Development. *European Journal of Social Theory, 4*(4), 427–442.
Trigilia, C. (2005). *Sviluppo locale. Un progetto per l'Italia.* Roma and Bari: Laterza.
Trigilia, C. (2012). *Non c'è Nord senza Sud. Perché la crescita dell'Italia si decide nel Mezzogiorno.* Bologna: Il Mulino.
Turiello, P. (1974). Avvento della sinistra al governo (1876) e sviluppo delle clientele politiche. In L. Graziano (Ed.), *Clientelismo e mutamento politico* (pp. 279–287). Milano: FrancoAngeli.

Vannucci, A. (2009). The Controversial Legacy of Mani Pulite: A Critical Analysis of Italian Corruption and Anti-Corruption Policie. *Bulletin of Italian Politics, 1*(2), 233–264.

Vannucci, A. (2016). The "Clean Hands" (Mani Pulite) Inquiry on Corruption and Its Effects on the Italian Political System. *Em Debate, Belo Horizonte, 8*(2), 62–68.

Vercellone, C. (2010). The Crisis of the Law of Value and the Becoming-Rent of Profit. In A. Fumagalli & S. Mezzadra (Eds.), *Crisis in the Global Economy* (pp. 85–118). Los Angeles, CA: Semiotexte.

Viesti, G. (2003). *Abolire il Mezzogiorno*. Bari and Roma: Laterza.

Viesti, G. (2009). *Mezzogiorno a tradimento*. Bari and Roma: Laterza.

Viesti, G. (2011). Le politiche di sviluppo del Mezzogiorno negli ultimi venti anni: scelte e risultati. *Economia e Politica Industriale, 38*(4), 95–137.

Weber, M. (1906 [2008]). Gesammelte Augsatze zur Wissenschaftslehre [The Logic of Historical Explanation]. Reprinted in Runciman, W. G. (Ed.). *Weber: Selections in Translation* (23rd ed.). New York: Cambridge University Press.

Weber, M. (1907 [1977]). *Critique of Stammler*. New York: Free Press.

Wolf, E. R. (1955). Types of Latin American Paesantry: a Preliminary Discussion. *American Anthropologist, 57*, 452–471.

Wolf, E. R. (1966). Kinship, Friendship, and Patron-Client Relations in Complex Societies. In M. Banton (Ed.), *The Social Anthropology of Complex Societies* (pp. 1–22). London and New York: Tavistock Publ.

Zitara, N. (1971). *L'Unità d'Italia: nascita di una colonia*. Milano: Jaca Book.

Zitara, N. (2011). *L'invenzione del mezzogiorno. Una storia finanziaria*. Milano: Jaca Book.

Zukin, S. (2010). *Naked City: The Death and Life of Authentic Urban Places*. Oxford and New York: Oxford University Press.

CHAPTER 4

Messina, From the Earthquake to the Present

4.1 The Roots of a Speculative Economy: State, Market, and the Messina Community

As discussed in Chap. 3, Messina is often stereotyped as a city "blocked" by the event of the earthquake and therefore unable to look forward to the future. Within the frame of the "infinite" rehabilitation and restoration plans, the lasting presence of the post-earthquake shacks, which still exist hundred years after the event, have become the metaphor for a static and immobile South that is unable to develop, and destined to remain backward. This is an ideological version that denies the city's history and aims to find a scapegoat for its "failure to develop" and for its "backwardness", devolving the blame onto the inhabitants dehumanized by the earthquake, who have erased its history and culture (Dickie 2000, 2002; Dickie and Sayer 2005).

On the contrary, we argue that Messina and its reconstruction are the expression of a capitalist development model, based on building speculation and property income, which had already gained a foothold in the first few years after the earthquake, but which was well established before. Throughout the whole of the twentieth century, and up to the present, the city has continued to expand its housing plans, following the demands justified by the organizational myths of "restoration", "economic development", "demographic growth", and "social need", but which, in reality, represented a means for enhancing the political and economic capital of certain social groups within the city's fabric. In this chapter, the aim is

D. Farinella, P. Saitta, *The Endless Reconstruction and Modern Disasters*,
https://doi.org/10.1007/978-3-030-19361-4_4

to show how this development model became established and how it changed during the different political seasons, adapting itself according to external stimuli, by changing the equilibrium supporting it, and following a dynamic that, as Polanyi says (1944), intertwines "state", "market", and "community" in particular and unexpected ways.

On the one hand, the territory is manipulated, produced, and reproduced by numerous public policies; in other words, the succession of municipal, regional, and national laws; development plans, programs, and variants; and also the omissions, absences, and delays have defined the face of the city over the years. In the eyes of the public institutions (Scott 1998), the territory is periodically defined as being in need of "restoration", "expansion", "completion", and "public housing", through political choices that underlie visions of the city that are sometimes implicit, or more explicit. These are translated into technical laws that are presented as neutral (moving the population, changing the designated use of land from agriculture to construction, expanding building to cope with population growth, locating poorer neighborhoods in certain areas, etc.) but have real and political effects on the people enduring them and on the territory. As already observed, the territory is also shaped by the choice to take no action and the failure to implement policies. For example, in those areas not covered by the Borzì master plan because they were not included at that time, the failure to approve a new development plan, which lasted until 1978, resulted in wild urban sprawl. Again, the failure to implement the new plan approved in 2002 has exacerbated the lack of essential services and of primary and secondary urbanization, to the city's detriment.

On the other hand, the territory is at the center of a speculative economic model, with the property market and construction at its core. From being common property, the territory becomes a fictitious commodity (Polanyi 1944), and its private consumption is destined to become a tragedy of the commons (Hardin 1968). Under this market action, the city becomes a hopeless place, oppressed by the seemingly irreversible deterioration of the urban and environmental fabric, and the object of unrestrained and chaotic urbanization that aims to "construct buildings" and create dormitory suburbs, without any thought for the creation of shared collective spaces with facilities, green spaces, and infrastructures. In particular, the building industry has moved into the hills behind the city, where agricultural activities have been abandoned (these supported the land with terracing, and grapevine, olive, and citrus cultivation), thus creating dramatic changes in land use which have caused landslides and subsidence during heavy rains.

Finally, situated between public action and the market is the community, that group of people who live in the city and its neighborhoods. In Chaps. 5, 6, and 7, we have concentrated on the inhabitants of the poorer neighborhoods, who live their daily lives moving around in the circular system that connects formal and informal work, social housing, and "casette ultra-popolari" (i.e., masonry houses, single room with no services or foundations, initially designed to temporarily house the poor in the post-disaster phase, but which, over the years, have been extended by their inhabitants), old and new, legal or illegal shacks. It is a layered and segmented community, which, in the face of so much state and market regulation, tries to organize ways to contrast this, by opposing, exploiting, or simply by ignoring it. Each social actor plays the cards in their hand, to affirm their existence, and loudly assert their presence in the modern world—a presence that has been symbolically denied by public action, by their removal from the center and relegation to the margins and more "hidden" areas of the city. As is seen in Chap. 6, by using certain specific tactics (De Certeau 1990), these subjects create claims of an ever-increasing complexity. These can vary from asserting their presence (in a sort of Heideggerian *dasein* or rather, given the southern context, in a De Martinian *esserci*—being there),[1] to the re-appropriation of spaces, using self-organization and alternative and informal forms of regulation that contrast with institutional forms. They will even pursue forms of social mobility, through their conquest of the right to public housing.

Throughout this chapter, we intend to focus on how the "State" (more generally understood as "public action") and "market" consolidate within a capitalist development model based on speculative building.

As underlined in Chap. 2, the earthquake has acted as an "accelerator" of ongoing social processes, which, in turn, trigger new economic, social, and political dynamics. On one side, the earthquake accentuated the city's decline as a commercial hub and interface between the city and the countryside, between agriculture and industry, due to the crisis in the citrus cultivation sector. On the other, at a time when there was a decrease in the investment opportunities linked to the processing and exportation of citrus fruits and derivatives, that had been historically characterized by speculation and intermediation (Barone 1982), the earthquake became the source of new opportunities for investment and speculation. Property became the most profitable investment sector, partly in response to the new reconstruction legislation and partly in response to the huge amounts of money that were to be invested by the State (Barone 1982, p. 2009).

Public institutions and private businesses joined forces, supported each other and became complementary, in a production model that produced both economic profit and political consensus, but offloaded the costs onto the territory through opportunistic consumption. It was a model which, despite being strictly appropriative in nature, was also able to redistribute during a good season, even to the more peripheral social circles, guaranteeing particularistic inclusion mechanisms also for the poorer classes. These mechanisms—especially for those able to take advantage of clientelistic channels—could put subjects on the path toward social mobility.

As anticipated in Chap. 3, a model such as this works and is socially legitimate, and even considered "virtuous", if the clientelistic mechanisms generate social inclusion, even particularistic ones.

4.2 Proto-financialization and Decadence

That the Messina memory is a bourgeois one, tempered with a precise notion of decadence, is an impression that stems, among other things, from the absence of the working-class subject from the scholarly literature devoted to the pre-earthquake city. Indeed, we know much about the socialist movement and the working-class formations of the city, but we know almost nothing about the everyday lives of those who occupied the lower rungs of the local social ladder. With regard to the pre-earthquake era, local historiography is very much concerned with the life of the institutions, and with the many characters that shaped the local intellectual and commercial bourgeoisie. The working class and the dispossessed appear mostly in the post-disaster literature, when these classes get to shape the new city and to characterize it in the terms of a social question. But even in this case, they are mostly a collective and obscure subject—that is, a mass that does vote for one party or another, that follows one patron or another, or that lives in shanties. Intimate portraits of these *popolani* and the studies of their daily lives are, in sum, quite rare. Our reconstruction, then, is affected by the lack of detailed information on this population. Fortunately, however, we do know much about the formation process of this significant component of the new city life. A process that rests on the dynamics set in motion by its counterpart, that is, the local bourgeoisie and the national parties.

Then, an investigation of those "underclasses" that are, until today, at the very center of the city organization has to start from the beginning.

That is, from the economic and political choices that shaped the city the way they did.

With regard to the economic conditions of the city over the years that preceded the seismic event of 1908, as it was shown among many others by Battaglia (2003, p. 111 passim), at the turn of the twentieth century, the citron industry had become the main industry in the city and the island (see Chap. 2). Although exports included other valuable goods, at the beginning of the new century, exported goods were ever less differentiated. Notwithstanding that, over the years 1898–1908, the cargo volume handled in the city port amounted to 4,135,000 tons (Battaglia 2003, p. 129)—enough to make Messina's port compete with the other big harbors of Italy. The city, moreover, was home to an international bourgeoisie of traders and industrialists. Since the early 1800s, Messina had received flows of merchants, employees, technicians, and craftsmen who came from all over Europe—as shown by the so-called British Cemetery, which, in spite of its name, still hosts in relevant numbers the remains of citizens from Germany, Switzerland, Denmark, Sweden, Greece, Russia, and, of course, England (this latter, the most significant group in numeric terms) (D'Angelo 1995). Since the free port had been set up in 1825, Messina had been home to about thirty British merchant companies. In comparison, Naples, a much bigger city, hosted only nine such British merchant companies (Battaglia 2003, p. 67). However, over the course of the time, many different international companies and private traders added to this original nucleus. Carbide, iron, sulfur, wine, fruit, and maritime insurance were some of the sectors within which these international traders were active.

But on the internal front, in spite of the international presence and the wealth of commercial exchanges taking place in the city, Messina was facing a crisis. Since the Unity, in 1861, the local bourgeoisie had tried to enter the new national production markets. But the attempts by the southern entrepreneurs to join the emerging industrial landscape had always been rejected, as the island represented for the new unitary government mostly an outlet for the goods produced in the north of the country. Within the national labor division, the role of Messina as well as of the South had to be a commercial one, and not industrial. A prelude, moreover, to that early shift from primary industrial production activities to the service sector that would go on to characterize both the local and the southern development in the following decades (Singelmann 1978; Gagliani 2003; Farinella 2009). Not by chance, at the dawn of the twentieth century, the textile industry in Messina had decayed almost entirely. The silk textile industry employed no more than eight people. In 1896, the cotton industry was represented by only one company, which hired about forty workers. The metallurgy industry was comparatively stronger, but it employed less than 1500 workers. In connection with the technological changes of the time, in the early twentieth century, new enterprises related to the trade of oil and electricity were started. The main company active in this field, *Società italo-americana*, today Exxon Mobil, employed 130 people full-time and, in addition, hired tens of temporary workers during peak periods. With eighteen firms active in the ten years that preceded the earthquake, the chair industry was well developed. The biggest of these companies employed sixty workers. With seven companies operating in the city, in the same period, each of the firms active in the cement industry employed about 130 people. The main local food processing companies employed on a permanent basis more than 300 workers, to which one should add a mass of unskilled workers who did informal logistical work in that sector. Processed food, finally, was regularly exported to the USA, Austria, Hungary, Greece, Egypt, and Canada, among other nations. Also important were the barrel manufacturing industry and the leather sector, which employed several hundred people. To this, of course, one should add the workers employed in the fishing sector and in agriculture, which amounted to more than 14,000 workers. Overall, at the very end of the nineteenth century, the labor force active in industry was composed of about 39,800 workers. People employed in the service sector (mostly, commerce) and professionals such as lawyers and medical doctors amounted to about 12,000 (Barbera Cardillo 1981; Checco 1990, p. 77 passim; Battaglia 2003, p. 112 passim).

More detailed account of the structure of the job market, according to the 1901 census, is shown in Table 4.1.

What Table 4.1 suggests is that the city presented an economic organization that did not fit the changes that were taking place at a global level. Still involved in international exchange networks, but subjected to competition with other continental markets, at the beginning of the twentieth century, Messina showed the prevalence of traditional sectors and a backward industrial structure that did not allow its insertion into modern production. Already, in the pre-earthquake era, we can see some of the job market characteristics that would become even more prominent over the course of the twentieth century. More specifically, there was a high percentage of unemployed men (a good 20% of the total), a sign that the local job market did not create enough work. There were a consistent number of male workers employed in the "Agriculture and Fishing" (22.9%) and

Table 4.1 1901 census: workers for sector (> 9 y.o)

Sectors of employment	*Men*		*Women*	
	N	*%*	*N*	*%*
Agriculture and fishing	13,293	22.9	1219	2.0
Extractive industry	5	0.0	0	0.0
Food processing (agriculture and fish)	6732	11.6	955	1.6
Metallurgy	1461	2.5	6	0.0
Construction	8133	14.0	31	0.1
Textiles	772	1.3	1776	3.0
Chemical	82	0.1	5	0.0
Service	5205	9.0	203	0.3
Commerce	3168	5.5	1073	1.8
Liberal professions	7165	12.4	636	1.1
Other, unemployed, non-active	11,978	20.7	53,859	90.1
Total	57,994	100	59,763	100

Source: Our adaptation from Mortara on Italian Census 1901 (1913, p. 261)

"Food Processing" (11.6%) sectors, the latter representing the main local industry. When we consider that a large number of workers in the two sectors were employed in the harvesting and processing of lemons,[2] it is evident that the production structure was fragile, as it was non-diversified, and overly reliant on a sector in which the goods (lemons) were already highly commodified and subject to unstable trends in the international market. Furthermore, well over 14% of the male workforce was employed in construction, demonstrating that this was an important area of employment for the local population.

Rather, the encounter with modernity underlined the city bourgeoisie's predilection for urban speculation and finance. At the end of the nineteenth century, these social classes were already moving toward those sectors of capitalism that allowed them to generate profit without any real investment in the production-manufacturing sector, but only income through speculation. Moreover, the citrus fruit sector, organized through a complex system that included advance payments, loans, and "bets" on sales, had refined its financial mechanisms in view of the profit opportunities to be made through intermediaries, linked to the various passages of the goods on paper (Lupo 1987, pp. 103–110).

As De Pasquale (2006, p. 144 passim) points out, at the heart of the speculative management of city affairs was a business clique of landowners, building developers, and local politicians. This is an element of continuity with the post-earthquake period, in which the "modus operandi" of maximizing profit without investing in industrial or production infrastructures is consolidated.

These were adaptions to current changes in the national and international economy, and, of course, to the new political scenario emerging from the Unification of Italy. General changes that, as noticed by De Pasquale (2006, pp. 103–106), pushed local reformers to speak of plights, or, rather, "plagues" that afflicted the city at the end of the nineteenth century. While this language was quite common in the social and political discourse of the time, from Europe to the USA, from the European positivism to the North American reformers of the Progressive Era (Tyrrell 2010), the local forms of this discourse were, ultimately, a reactive response to the new territorial hierarchies produced by the Unity, rather than the mere expression of moral instances aimed at disciplining the poor. In the local context, in other words, this was a discourse produced by the bourgeoisie and addressed mostly to the members of this class—in spite of the typical concern of the time with those lower classes who did not accept their position in society (De Pasquale 2006, p. 106). In 1903, such "plagues" were represented by the national government that abolished the grain tax, and by the local municipality that did not know how to improve the conditions of the poor neighborhoods and villages surrounding the city, nor to prevent rents from skyrocketing, or the failures of several local banks. Not to forget the sanitary conditions of the poorest areas of the city and the growth of organized crime led by members of the local elites.

Among other things, these concerns were the outcome of the anti-southern policies implemented at the end of the nineteenth century by the government led by Pelloux and the weakness of the countermeasures undertaken by the prime ministers Saracco and Zanardelli at the turn of the century (Saladino 1966). Moreover, in those years, serious scandals involved important national banks. In 1889, the Banca Romana scandal, which involved businessmen as well as politicians and prime ministers, showed the entity and extent of corruption. By means of that bank, the national politicians had created a complex network aimed at financing and receiving support from the world of business. Moreover, besides the many financial crimes committed by the board, Banca Romana showed a number of bad debts due to speculative maneuvers in the field of real estate—something that had terrible effects on small account holders. Such uses of the bank system, however, were quite common and could be found in many parts of Italy, including Messina. Here, during the same years, the convergence between private interests and politics became clearer. Since the early 1890s, the city was ruled by a tiny circle comprising big landowners, real estate entrepreneurs, and financial brokers. In this context, the landed aristocracy, which had moved to the city, despite continuing to hold its old interests in the country, joined the urban circles of power. It was clear, in fact, that the building of roads, public buildings, aqueducts, railways, and other infrastructures was related to the land. In these circumstances, the options available to local authorities consisted of seizing the land or buying the rights of use by the landlords. As shown by the high employment levels (Table 4.1), the building sector represented an investment area within a growing speculative economy, thanks to the widespread conviction that the city at the end of the nineteenth century was an industrial and commercial center undergoing rapid growth, both demographically and economically, so that new infrastructures and new housing were essential. In fact, from the extension of the docks to the building of a new aqueduct and the sanitation of the city, Messina needed new infrastructure (Parrinello 2015, p. 48 passim). Although the option of buying rights of use was uneconomical and could create budgetary gaps, in most cases, this became the choice of the City Council, exposed to the influence of the agrarian bourgeoisie. The city, in fact, relied on the central government and its willingness to fund possible gaps resulting from similar choices (De Pasquale 2006, p. 145).

As far as housing was concerned, the stringent building constraints established after the 1783 earthquake (that had completely destroyed the city), which aimed to limit the number of buildings, their height, and ensure they were adequately distanced, were gradually relaxed and ignored. The city had seen a disorderly growth in the height of buildings, and many new structures were closer to each other than before. Moreover, thanks to the collusion between the developers and the political-administrative offices that granted building rights (De Pasquale 2006, p. 54 passim), many upper- and middle-class houses in the city center had large exterior decorations and balconies that made them heavier (which caused them to collapse during the 1908 earthquake). Other areas of speculative income for the city bourgeoisie came from the supply of the city's drinking water from private farmland in the surrounding countryside. The Council "on the one hand bought water from private owners and then distributed it directly to the city, but on the other, it conceded special rights to the larger landowners, and then bought it back in the event of a shortage" (De Pasquale 2006, p. 71).

The ultimate outcome of this way of managing public interests was the strengthening of ties between sectors of the old aristocracy and the most entrepreneurial middle classes—the meeting, in short, of old and new forms of capital. Over the course of a few decades, that encounter determined the creation of six local banks and the raising of a new generation of adventurous brokers (De Pasquale 1992). These figures, which acted as intermediaries and consultants, collected the savings of big and small depositors and invested them in dead-end projects. Thus, several bubbles burst at the turn of the new century, and this was how local banks, controlled by a handful of businessmen, amassed large buildings and land stocks. One example of such speculations is the implementation of a tramway, *La messinese*. In this case, a company was created and funds collected; finally, the bank that supported the initiative was soon liquidated, along with the associates who lost lands, bonds, debt securities, and capital. The purpose of this malicious investment offer consisted in embezzling the land of big owners in order to sell the rights of way and the spring waters to the municipality. Bubbles and frauds, thus, were part of the war launched by a new emerging urban middle class against the agrarian bourgeoisie (De Pasquale 2006, pp. 150–151). But they also constituted a threat to the public purse—in view, above all, of the new infrastructural plans that promised to change the features of the city. These complex plans—which have been aptly described by Parrinello (2015)—resembled other similar urban projects in Europe. Plans that were influenced by recent tendencies in the disciplines of public health and were aimed at sanitizing the most dilapidated neighborhoods (within which cholera and other diseases had become common), at enhancing the water distribution, as well as at modernizing the port and changing the course and the beds of city rivers. However, what is relevant for our purpose is to observe how Messina was integrated into a larger framework. It is important, in other words, to reflect on the modes through which the urban elites mirrored a social order and an organization of interests that had been created "above"—at the central level of government—and that, from there, had trickled down and reached the dispersed territories of the nation. Yet, it is relevant to notice how, already at the turn of the century, and even in this semi-peripheral space (see Chap. 2), the decadence of the manufacturing industry led, on the one hand, toward a regime of proto-financialization and, on the other, toward classical forms of parasitic rent that were translated into the urban environment. In Braudel's famous words (1984, p. 246), the phase of expansion of the immaterial in the capitalist cycle is "a sign of autumn". According to Chesnais (1996, p. 35), the same phase marks the "dictatorship of creditors", that is, the dominance of rentiers in mature capitalist economies.

According to David Harvey (2010, p. 40), industrial or production capital used to produce a commodity to be sold on the market for a profit, is only the form of capital circulation that has come to dominate since the mid-eighteenth century. The central aspect of capital is not the production of goods, but the need to increase its value:

> Capital is not a thing but a process in which money is perpetually sent in search of more money. Capitalists—those who set this process in motion—take on many different personae. Finance capitalists look to make more money by lending to others in return for interest. Merchant capitalists buy cheap and sell dear. Landlords collect rent because the land and properties they own are scarce resources. Rentiers make money from royalties and intellectual property rights. Asset traders swap titles (to stocks and shares for example), debts and contracts (including insurance) for a profit. Even the state can act like a capitalist, as, for example, when it uses tax revenues to invest in infrastructures that stimulate growth and generate even more tax revenues. (Harvey 2010, p. 40)

In this sense, the process of financialization, in which capital generates profit in the simplified form M→M', instead of M→C→M', is not an expression of "mature" capitalism, but of a strategy for capital appreciation, which was increasingly pursued in response to the erosion of global profit margins after a brief revival in the 1980s (Harvey 2010, p. 28).[3] Although it is not yet possible to speak of financialization in the case of Messina, but only of forms of proto-financialization, its reorganization of capitalism leans toward economic forms of a speculative type, not for the production of goods, but for profit production. This undoubtedly represents an example of that ability to reconfigure social life, spatially and temporally, as a function of the profit creation of contemporary capitalism (Harvey 2010, p. 42).

Certainly, as observed by Lapavitsas (2013, p. 799), there are similarities but also important differences between the current notions of financialization and the earlier signs of financial ascendency in advanced capitalism. In the early stages of financialization—the process, in sum, that was witnessed in Messina at the beginning of the past century—the transformation of capitalism consisted of a mixture of industrial and banking capital. A process that led finance capital to organize the economy to suit its own interests, in guises that were both general and local. Ways that, in the case at hand, depended on, and were the outcome of, the coexistence of three types of capital related, respectively, to industry and trade, finance capital, and land. An interplay, however, that followed both models of informal regulation and negotiation sketched elsewhere—as we said, at the central level of government—and local models, which depended on power relations as well as on the individual and collective capability of "playing the game".

The role of political regulation becomes particularly important, as does the ability to take advantage of it, by binding it to the speculative interests of the business classes, a connection which, as we see over the course of this chapter, would come to constitute a social bloc for the control of the city.

With regard to these latter aspects, in 1902, only a few years prior to the earthquake, in a report addressed to the then-Minister of the Interior Giovanni Giolitti, the Prefect of Messina wrote that there was a monarchical group that plotted against the local branch of the party of government. It was a group, the prefect noted, of "carpetbaggers who did not accept their defeat in the aqueduct issue". In response, Giolitti ordered the Prefect to be cautious and not let any "old carpetbagger" enter the Provincial Council (De Pasquale 2006, pp. 166–167). Indeed, that monarchical group was composed of former members of the Catholic party, former socialists, and republicans. Given the Italian political scenario of the time, it was, in other words, a not so peculiar formation of monarchists, clericals, and liberals, which resembled similar politics seen in action at the national level in the years of the so-called transformism—this being the art of forming parliamentary majorities based on single personalities and politically contingent programs, which overcome the traditional ideological distinctions. This group, however, obtained the majority over the course of the next administrative elections. A year of protests and riots, thus, followed. In consideration of the new scenario present in the city and the national election of November 1903, Giolitti, now prime minister, sought new allies in Messina and, therefore, put together a composite group which included conservative Catholics who had opposed his faction in the past. The political history of the city until the seismic event of 1908, then, was one of endless political crises—a condition, however, which reflected the unstable general political scenario of the time, and that Messina shared with other cities.

4.3 Reconstructing Inner Circles: Geographies of Separation and Exclusion

On December 28, 1908, the earthquake struck. As Bevilacqua (1981, p. 189) has noted, disasters seem to produce a "galvanization of power", which stresses the centrality of the State (Noto 2008, p. 199) and produces, at the same time, a surplus of destruction. That is, the "violence of delayed destruction", which is incremental and the repercussions of which are felt across a range of temporal scales (Nixon 2011, p. 2; Iovine 2016, p. 85). A temporal scale that, with regard to the chronology of the basic reconstruction plan, was spread out over almost two decades, or, from our perspective, over a full century. The concept of reconstruction, in fact, does not relate to space only. Beyond the restoration of the "physical"—houses, infrastructures, or public buildings—the concept of reconstruction should also imply the ideas of everyday and social order. States' views of post-disasters tend to be performative and focus on those quantifiable elements that measure the efficiency of state interventions: for example, the number of people who were rescued and housed, or the number of buildings that were restored in a given amount of time since the beginning of the crisis, and other similar indicators that have the characteristic of being easily communicable and understood by the general audience. But there is another level involved in the practices of reconstruction that pertains to those elements that are not as measurable, and that imply the "Self" of individuals and communities. This level includes ways of being in the space and perceiving it; rituals and embodiment of social order, comprised the notions of social class and hierarchies (for a review of these points, Adler et al. 1987). Reconstruction processes, thus, are not aimed only at restoring space but also at re-establishing or constructing practices and public cultures, that is, those networks of collective and individual meanings and positions that compose the social world (people and institutions). In the case at hand, how "to work the system", who is the best patron, or how to simulate a social condition in order to fit the requirements for public benefits and so have access to the beneficiaries' lists prepared by such patrons are practices and wisdoms that can be generated, learned, and transmitted over the course of a reconstruction process. Certainly, like in the case of Messina, similar practices and visions might preexist and be part of the pre-disaster order. The calamity, then, marks only the continuity of a given public culture and confirms the importance of path dependence as a concept even in disaster scenarios (Ghimire 2018). In the words of Rixen and Viola (2014, p. 10), the concept of path dependence "captures a particular dynamic of reproduction over time in which the equilibrium is deepened [...] As it reproduces itself the institution becomes *increasingly* stable, locked-in, and resistant to reversal". Although there might be the risk of "stretching" this concept and so lose clarity and precision, it is clear that the poetic of change related to catastrophic events is, ultimately, vain. Especially in the public and economic discourse, on the traces of a certain vulgarization of Schumpeter's work, disasters are often seen as "creative" events that can generate economic and societal growth (for a review of the economic literature, especially in the field of technologies, see: Cuaresma et al. 2008). Space also is subjected to this popular association between disaster, rebirth, and creativity (Gotham and Greenberg 2014). Spaces and ruins can be seen as "constellations" of meanings and uses in opposition to the homogenous, empty time embodied by the classical ideologies of progress (Gordillo 2014, p. 20). A point of departure, in sum, for "something new"—whether the invention of heritage or the "smart cities" (two common and modern ways to produce an internal innovation of the classic discourse on progress as a linear process). If these common ways of understanding disasters are apt to describe what happens on the surface—that is, in the tangible points where visible manifestations of modernity inscribe themselves (e.g., the space)—they are not as effective in grasping the substance of social relations that persist under the new façades. Thus, while it is certainly true that space implies social relations, and that both new and canceled spaces bring about different social relations (Jacobs 1961; Holston 1989; Lefebvre 1991), it is also true that there are relational aspects that survive changes in space. In this sense, the deep ideology and the political praxes of the time within which disasters and dramatic changes take place overwhelm both physical alterations of space and urbanism as an intellectual and voluntarist practice oriented to goals. Urban change and "creativity" applied to space, therefore, at least in some cases, should be understood as blankets, or masks, that hide the continuity and latency of social structures.

Messina's case in the aftermath of the 1908 earthquake confirms these impressions. Already in Chap. 2, we have described how Prime Minister Giovanni Giolitti used the event to take control of the city by allocating resources, creating ad hoc financial agencies, granting loans to owners of damaged houses (in particular, the big owners), and closing both eyes to the many distorted uses of these measures. In particular, in 1910, Prime Minister Giolitti created the *Unione Messinese dei Danneggiati dal Terremoto* (The Union for those damaged by the Messina earthquake). The agency's task was to bring together the owners of damaged houses and coordinate the restoration or reconstruction of the buildings, which were to be built on the land controlled by the agency itself. Members of the Union would qualify for loans that could be repaid over thirty years, for a value corresponding to 75% of the value of their old house. As we have seen, this measure introduced a distinction between those who did not own a house before the tragedy or did not possess the means to repay their loans, and others, who were mostly the more privileged members of the local bourgeoisie. Moreover, the value of the loan was directly proportional to the value of the damaged house, and thus favored the wealthier owners. Finally, the loans were transferrable, and this paved the way for large-scale speculation—something that placed both the lower and middle classes at risk (Noto 2008, pp. 228–230), so that in order to counteract these effects, they organized themselves into cooperatives. In effect, the measure resulted in many survivors being denied the right to housing. Within this framework, those who found themselves excluded were forced to live in the barracks[4] that were to become a permanent feature of the urban landscape.

All these post-earthquake political measures were at the basis of the new social differentiation within the city, respectively, based on the rise of a tiny class of privileged beneficiaries, and of a much larger class of people who were excluded by these privileges.

The "capitalist use of the earthquake" (Barone 1982) is therefore at the basis of the current urban social stratification and, as described in the previous chapters, it is also one of the outcomes of that convergence of interests and alliances pursued by Giolitti, with the aim of stabilizing his precarious hegemony in the city.

Furthermore, since the earthquake in Lisbon in 1755, modern disaster interventions have generally consisted of the construction of temporary wooden houses, or barracks. In January 1909, the authorities selected sections of the city in which to build about 7600 temporary wooden shelters ("barracks"), which covered a total of two million square meters and were able to house around 60,000 people (Noto 2008, p. 209; Di Leo and La Spada 1985, p. 12). Along the Giostra torrent, 2093 such shelters were erected, and in the Moselle area, in the center of the city, a further 3000 shelters were added. In the interior areas of the city that were free from rubble, a further 400 wooden houses were built, while near the Zaera torrent, these amounted to about 2100. In 1911, when the first post-earthquake development plan—the so-called Borzì master plan—was approved, there were wooden barracks in every part of the city (Di Leo and La Spada 1985, p. 12). The principle that led to the selection of the areas consisted in the identification and anticipation of the possible development lines that the next master plan would have followed. In many cases, these choices would be a prelude to the explicit and institutionalized geographies of exclusion that would have characterized the city for over a century, in a stratifying logic that mixed, or separated, different types of houses and social classes.

In Sibley's terms, in fact, the city is a landscape of exclusion. Some expulsion processes are opaque and visible only to those who are expelled from the center of a community (the space, for instance, denied to certain ethnic groups and classes by means of security measures, or the access costs to services and goods); others are at the very basis of the structural process of formation of the cities (Sibley 1995, ix). This latter case being confirmed by the class-biased nature of the first development plan, the Borzì master plan of 1911.

The Borzì plan[5] envisaged the demolition of the barracks erected in the center of the ancient town and the subsequent reconstruction of the new city on a grid plan, with long broad arterial roads (at least ten meters wide), which would be intersected at right angles. The buildings were to be situated in the resulting rectangles, each with a large internal courtyard with a maximum of two floors, and they were not to exceed eleven meters in height. It was a rational plan, in line with other urban designs established in Europe at the end of the nineteenth century, and it was inspired by the functionality of the Roman *castrum* model, which allowed rapid connections and wide escape routes. It was a reference to the aesthetics of the garden city, for its characteristic expansion in length rather than height (linked to the necessity to obey the new anti-seismic laws), the long tree-lined streets running through the center, and the urban green spaces created inside the large internal courtyards.

The master plan contemplated an administrative and commercial section close to the port, and an industrial area, which was also near the sea, along Via La Farina. The northern area was destined for upscale residences—mostly mansions and luxury houses.

The class-biased nature of this plan was evident because there was no reference to public housing and projects, and the plan concentrated on services and public buildings in the city center only—that is, in the same urban space where, soon, residential neighborhoods and houses for the upper classes would be built. The Borzì plan, in fact, ignored the more external areas, which, in view of the city's demographic growth, were destined to become expansion zones for the lower classes. These lines of expansion extended to areas which, up to that moment, had been peripheral, mainly agricultural lands to the North of the city (*Torrente* Giostra-Annunziata),[6] the South (*Torrente* Gazzi), and the Center (the higher parts of *Torrente* Zaera) (see Fig. 4.4 in this chapter). The plan was to move the people who had been temporarily housed in the city center to social housing in these peripheral areas, along the descent lines of the ancient "torrents" that flowed down from the Peloritani hills through the city to the sea. The new public housing was to be located at the base of these riverbeds, in the internal flat valleys, where there had once been rural villages.

On the one hand, the choice was directly connected to the location of the first post-earthquake camps, and it is sufficient to remember here that in 1909 almost 10,000 people had lost their homes and were living in makeshift huts set up in these areas external to the city center, thus creating shantytowns in the villages of Annunziata and Fondo De Pasquale (northern zone), Fondo Fucile (southern area), and Camaro (central area, *Torrente* Zaera).

On the other hand, the choice reflected the desire to rebuild the rural villages that had once been connected to the city (Lupo 1987, p. 94) and were the expression of the dynamic relationship between the city and the countryside typical of city development models in the pre-earthquake era (see Chap. 2; Barone 1982). In fact, the same plan envisaged a large industrial area in the southern part of the city, dedicated to citrus processing, which would be connected to the new railway station that was to be built next to the port.

In addition to being classist in design, there were many other shortcomings to Borzì's plan. He did not take into account, for example, that the development model based on the exportation of fresh citrus fruits and citrus derivatives was already in decline (Cicciò 2016; Battaglia 2003). At the root of the exportation and intensive agricultural crisis were the customs tariffs imposed under Giolitti, and which continued during the fascist period, causing the gradual abandonment of the countryside in favor of the city for an ever-growing number of the rural population. The interior areas below the torrents, between the hills and the city, would no longer comprise rural farming villages, and deprived of this productive-functional role and without a development plan to limit growth, they would gradually become peripheral pockets for the popular urbanization of the city's fabric, in reply to the increase in demographic pressure. In turn, this generated the progressive urbanization and consumption of agricultural land, which tended to move higher up into the hills, causing serious hydrogeological instability. The plan, thus, concentrated the lower classes far from the city center, in areas that lacked minimal services and were also exposed to landslides and other environmental risks (Caldo and Santalucia 1997, pp. 109–110; Ginatempo 1976, p. 182 passim; Noto 2008, p. 220).

In spite of the fact that there are other, more benign interpretations of the Borzì master plan (Lo Curzio 1985), it should be out of question that, regardless of the intentions of the proponent, this plan paved the way to future speculations and, implicitly, to radical class divisions. The limited height of the buildings (ten meters in the original plan; fourteen meters in a 1935 revision) required an extension of the building land and the inclusion of a large number of landowners in the circle of the stakeholders. In the long run, windfall allocation of resources from the central government affected the formation and distribution of incomes. As we have already said, the profile of the local elites was conditioned from these new business perspectives, so that real estate investments replaced the old maritime trades—a change, however, that was already taking place in that highly speculative and proto-financial framework that preceded the seismic crisis, and that this crisis was only making extremely visible. A new direction, however, that depended also on the capital flight determined by the destruction of the city and the lack of clarity on the future of the place. As a matter of fact, the State did not support local industries and business by reactivating production facilities and encouraging big firms to continue their operations in the territory of the city. The 1911 census, thus, portrayed already a city whose economy was mostly dependent on the State, in the form of salaries for the personnel of the public administration and pensions. Public sector, followed by craftsmanship and small retailing, were now the core sectors of the city (D'Angelo 2009, p. 100).

In the years after the disaster, thanks in particular to the opportunities provided by the construction industry, the city population had grown from 127,000 people to 176,000 people. Yet, in spite of the growing demography, the reconstruction was slow. If the 1911 Borzì master plan envisaged the building of 28,000 houses in twenty years, in the early 1920s, only 1482 of such lodgings were available (Noto 2008, p. 239).

The presence of the barracks posed problems and dilemmas related to the implementation of the development plan. On the one hand, they occupied much of the land; on the other, as most land property deeds had been lost, land property needed to be protected, and the barracks were apt to pursue this objective. Such concern for property was not ideological but related to the political plan being pursued by the government, entirely centered on the stakeholders' interests. Therefore, the idea of simply seizing the land on which the barracks were located was rejected almost immediately.

In order to overcome these difficulties, the authorities set up a two-step plan. During the first phase, the City Council would acquire the property of the barracks and the land on which they were situated. In the second phase, it would concede the opportunity to form landowners' consortiums by setting up the Union mentioned earlier. This agency, however, would be very slow in starting, and further delays would allow large consortiums of landowners to flank and replace the Union for major operations and conduct the works with a certain degree of independence. However, in 1914, the agency split into two parts, and it was complemented by the *Ente Edilizio*, the Housing Agency, which became responsible for public housing. Nine years after its constitution, the old *Unione* had built 12 blocks and 177 residential buildings, 56% of which were 6-room apartments for the middle classes, while 32% of this housing stock was made up of upper-class apartments and 12% of luxury houses (villas and the like). During the same period, the Ente Edilizio had built 7 blocks and 196 houses for State employees and 437 *case ultrapopolari* (Di Leo and La Spada 1985, p. 14). The reconstruction, thus, proceeded at a very slow pace, especially considering that between 1911 and 1921, the population of the city was constantly growing, until it reached about 177,000 (as can be seen from the 1920 census). The delay was also due to the outbreak of the First World War. In 1919, a year after the end of the conflict, the rehabilitation process was at a standstill, and the first wooden barracks that had been erected the day after the earthquake were becoming increasingly dilapidated.

As Longo pointed out (1933, p. 188 passim):

> [Messina seemed] not a city but a camp of old, soaking-wet shanties [...] After 12 years under the constant action of the rain, sun and wind, there had been a shocking amount of deterioration: roofs, walls and floors were rapidly disintegrating. The pine and spruce foundations that held up the floors were rotten and stood a little askew on the ground. Earlier, an indiscrete comment had termed the shacks "doghouses". But by 1919 they had become very much worse than the doghouses a hunter with even a minimum amount of respect for his hounds might possess.

This is how some of the characteristics typical of the southern housing model (lack of housing, overcrowding, illegal building in response to the lack of housing) came into being in the post-war period. They also go some way toward explaining both the building boom of the 1970s and the high level of tolerance toward illegal building practices (Bellicini 1990; Cremaschi 1990; Gucciardo 1996).

To this social and economic basis, finally, the mass of the unemployed was added—ever more so after the end of the First World War. The social landscape of the modern city, then, was already visible and delineated three years after the earthquake. In short, this new organization and the related class divides posited the difference between those who were guaranteed and those who were not.

4.4 Fascism and the Consolidation of the Shacks

Within this framework, in the 1920s, the advent of Fascism brought about truly important changes. If the slow pace of reconstruction until that moment was the outcome of the prevalence of particular interests and the result of the alliances between local and national powers, the rise of the new regime upset this equilibrium.

First, Mussolini replaced the heads of various branches of public administration, including the reconstruction agencies that had been under the control of political and economic cliques opposed to him (Baglio 2005, p. 15 passim).

Second, Mussolini strengthened his relationship with large construction firms that were to carry out many of the projects for the Fascist regime, with the aim of changing the features of many Italian cities for speculative, political, and celebrative reasons (Bortolotti 1978, p. 153). The same process and alliances between State and capital that, among other things, would bring about the so-called *sventramenti*, or demolitions, of Rome's historical center and the subsequent emergence of the *borgate*—that is, the shanty towns at the margins of the Italian capital. Third, discontent and unrest were spreading in Messina. Thus, if the repression of these mobilizations was the stick, the acceleration of the reconstruction process was the carrot.

In June 1923, Mussolini arrived in Messina and announced a new plan for the city. To signal a break with the past and the start of the rehabilitation process, Mussolini dissolved the *Unione* and entrusted the restoration to Civil Engineers, who were part of the national state administration. In Messina, among other things, he met Bishop Angelo Paino. This was the first in a long series of encounters with the priest, which, over the years, transformed him and the local church into the most important realtors in the city, up to the point that the pastor became Mussolini's most important political ally in his struggle against local opposition (Noto 2008, p. 50 passim).

Mussolini also announced his intention to create all the necessary infrastructures for the economic development of the city (e.g., the new central station near the port), to encourage clearance of the barracks and to respond to the general lack of housing which characterized the post-earthquake city, by supporting private and public construction.

Moreover, it is with the rise of Fascism that the reconstruction process began to accelerate and with it the inherent classist character of this process and the capitalistic use of the earthquake.

Supported by public spending linked to the national funds for the rehabilitation process, private capital was increasingly moving from the citrus fruit sector to construction. Anonymous business companies involved in financial intermediation, construction, and service distribution (electricity, gas, and transport) (Cicciò 2016, p. 93) were set up. In 1935, pressure from these new investors led to the decision to waive the Borzì master plan, allowing buildings to reach a maximum height of sixteen meters and four floors above ground level. This meant that housing capacity was doubled in the construction areas (Budà 2012), but it was also the beginning of widespread private building speculation in the city center, in the race to add extra floors to already existing buildings.

The development of the city was divided between central and external areas; the houses and various zones were organized according to the inhabitants' professions or activities. In the wake of the Borzì plan, the city center mainly consisted of private upper-class housing.[7] In proximity to the center, sometimes alternate to private households, there were the so-called "economic houses", built by cooperatives of landlords, or workers belonging to mutual organizations.[8] Yet, the houses for the public servants (built under a special housing program for State employees, and endowing this category with much better houses than the rest of the population), with the exception of the railway workers, were placed in different areas of the center, but mostly close to the cheaper housing.[9] The public projects, moreover, were initially concentrated in distinct areas located both north and south of the center.[10] Finally, destined to the lumpen, there were the so-called *case ultrapopolari*, that is, the ultra-poor and neglected masonry housing placed in remote areas or in the hidden, rarely visited, locations inside the city limits.[11]

In the 1930s, the barracks constituted 31% of the available housing stock. In this respect, it is important to notice that the difference between the *case ultrapopolari* and the barracks was almost non-existent and mostly limited to the used materials: masonry and wood, respectively. Furthermore, they contained different facilities. For example, only 23% of the working-class tenements were provided with drinking water, in comparison to 73% of upper-class homes. In addition, only 55% of the poorer houses were provided with toilets. The crowding rates were 3.7% inhabitants per room, in a one-room house, and 2.2 inhabitants per room, in a two-room house (Di Leo and La Spada 1985, pp. 14–16).

The disposition of the different types of houses and the distance from the geographical and social center—from three to five kilometers (a distance that was doomed to grow over the course of the decades)—show very clearly the ideology and the hierarchies that led the reconstruction of the city. This was perhaps a sort of global, latent ideology of the time, if one considers, for example, the "fight over the relocation" of Chinatown in post-earthquake San Francisco (Dyl 2017, p. 152 passim). Certainly, the outcome of reconstruction in Messina was—and still is, as shown in this chapter—a highly differentiated social space, which reflects local ways of producing environmental inequality, if not racism (Cutter 1995, p. 112). This latter being, ultimately, the exposition of social classes to different living conditions and health regimes. A form of inequality, moreover, that expresses instances of segregation and "coupling"; or, if you will, of tolerable proximities among classes—together with the deliberate introduction of different regimes of visibility and the production of desirable social landscapes (Brighenti 2010).

The start of the shack-clearance process (which, as is seen, produced the paradoxical effect of peripheral shack building) strengthened the urban geography of exclusion.

This intensive shack clearance mostly took place in the historic center of the city. By 1928, 8723 one-room barracks had been demolished, and by 1932, the number had risen to 13,971. The construction of popular housing continued in the areas outside the Borzì perimeter, and in 1927 alone, 2000 social houses were built in the Camaro and Gazzi areas.

However, the effort did not go far enough, even though 5492 housing blocks were finished (Di Leo 1985). According to the 1931 census, one-third of the population were still living in the existing 7676 barracks (32,408 individuals) (Longo 1933, p. 375). The housing census of 1936 revealed that over 35% of houses were overcrowded and that almost 49% of the population were living in them (Di Leo 1985).

The problem was that the shack clearance was taking place at a much faster pace than the new buildings. Moreover, in order to demolish the shacks, the city needed to expand, making it necessary to seize new land.

Mussolini therefore decided to build the *casette ultrapopolari* which were to serve as a stopgap for the urban underclass that had been "purged" from the city center. Obviously, these *casette ultrapopolari* were located in the most remote and peripheral areas, at the margins of the urban city, where already in 1909, some temporary accommodation had already been set up and where some of the popular housing was being built. These areas, represented by former agricultural lands, and known as "fondi", enlarged already existing sites, or created new ones in proximity to them. They were reserved for the construction of "casette ultrapopolari", while in the adjacent areas, there was a huge building project for the realization of popular housing. The initial plan was to use a rotation mechanism, whereby the inhabitants in the shacks and "casette ultrapopolari" would be moved directly into the new housing being built close by.

It was immediately evident that the new social housing was being built far too slowly, and there were insufficient houses in comparison to the number of barracks being demolished. The architectural panorama of large parts of the "marginal" territory situated next to the torrents, in the South, North, and Center, began to show signs of the disorderly coexistence of post-earthquake barracks, *casette ultrapopolari*, and social housing, interspersed with new informal and illegally built shacks. In this way, the current structure of the city started to take shape, but far from being the result of the earthquake, it was more the result of the distorted planning of the Fascist state, combined with the speculative interests of private businesses, that affected the historic center.

The construction of the "casette ultrapopolari" was seen as "window dressing", with paternalistic and moralistic overtones, and it involved the substitution of the wooden shacks in the city center, which had already become quite ramshackle, with brick housing erected in areas less visible to the eyes of the bourgeois classes, in the more peripheral and marginal areas of the rural suburbs.

The *Mussolini little houses* (le "casette di Mussolini") (still so-called by local residents) are single-story, one- or two-room single brick buildings, with cane roofs, and no toilet facilities. According to several residents in the Gazzi district, including a woman who was born there and was still living there in 2002 (the date of our informal interview), they were masonry shacks built directly onto the earth, without any solid foundations. Also, they were built using a hastily combined mixture of earth and cement to hold up the precarious brick structure, and not the reinforced concrete that had been declared the only legal building material for seismic areas after the 1908 earthquake. They were "temporary" dwellings, precisely because they were not suitable for a high-risk seismic area. The fact

Fig. 4.1 A portion of the "Fondo Fucile" shanty town seen from above. Social housing blocks of Villaggio Aldisio are also visible in the background. Source: Photo by D. Farinella, 2019

that they were temporary was also a justification for the lack of toilet facilities; it was necessary to make sacrifices while waiting for public housing.

Unfortunately, although initially only a temporary measure, over the years, the *fondi* have become consolidated areas, springing up alongside public housing built in the same area, but without any sense of continuity (Fig. 4.1). As already mentioned, this creates an untidy landscape, where blocks of public housing from different periods are interspersed with "casette ultrapopolari" in various states (some derelict, others, a great many, re-modernized, enlarged, with the addition of more rooms or floors), shacks, older, and more modern buildings (Figs. 4.2 and 4.3).

In 1960, a proposal for a national rehabilitation law (De Pasquale et al. 1960) identified the housing in the *fondi* as the main reason for the deterioration of the urban fabric, calling for its elimination.

Fig. 4.2 "Casette di Mussolini" in Gazzi, informally restored and overhead by the inhabitants. Source: Photo by D. Farinella, 2019

Fig. 4.3 "Casette di Mussolini" in Gazzi, informally restored by the inhabitants. Source: Photo by D. Farinella, 2019

The first problem that needs to be addressed by those concerned with the unfinished reconstruction of Messina is the rehabilitation of the so-called "Fondi". These are small houses, in a state of dilapidation, that were built on areas of public land as "temporary shelters", destined to be demolished as soon as cheap housing for the homeless had been built. The proof of this lies in the fact that these buildings were recently excluded from the right to

> buy, because they are destined for demolition. However, the new housing has never been built, and the most disadvantaged and poorest members of the population have been living for over 30 years in those *temporary shelters* lacking even the most basic hygiene requirements, and in increasingly serious conditions of overcrowding. [...] There are about 2,000 shelters in which around 5,000 families are forced to live, often with two or three families per shelter! (De Pasquale et al. 1960, p. 3)

From the extract, we can see that the number of *casette ultrapopolari* was around 2000, with 5000 resident families, so we can estimate the population as between 10,000 and 15,000 people (considering, on average, a family unit was made up of 2 or 3 people, although this is probably an underestimation). "Doghouses" is the word that Ginatempo (1976, p. 154) used to define the quality of this housing which, in the meantime, had been inherited by the descendants of the early tenants, becoming a stable presence in the urban landscape. The 1960 proposal included a mini-census with a table showing the year, the location of the *fondi* containing *casette ultrapopolari* (together with the initial four post-earthquake areas), their size, and the number of existing dwellings. It is interesting to compare this table with that from the census carried out by the Messina Council in 2002 as part of the Regional Law n. 4 dated May 15, 2002, "Measures for the acceleration and completion of the rehabilitation of the city of Messina" (Regione Sicilia 2002) and recently updated (Comune di Messina 2018).

As can be seen in Table 4.2, after more than seventy years, the areas for rehabilitation still correspond to the areas in which the *casette ultrapopolari* were built many years after the earthquake (from 13 to 30 years after). This should have been an ample amount of time to build other housing. Far from decreasing, between 1960 and 2002, the number of shacks increased, and, in 2002, a further area was added to the existing rehabilitation areas. The new area is close to the San Filippo torrent, a little further south and, from the post-war period on, was the site for the large-scale construction of public housing. With regard to the population, in these areas, the 2002 Council Census registered 3336 families (with an estimated 11,500 residents). In 2018, the Council Census update on the state of rehabilitation, made by the Housing Policies Department, stated that, in August 2018, there were still 2722 families living in the areas for rehabilitation, with a total of 9744 people (Messina Council), despite large-scale construction work in the popular areas.

Table 4.2 Comparison between restored areas over the years 1960, 2002, and 2018

Location of the torrent areas	*"Casette ultrapopolari" according to the 1960 legislative proposal*				*Council census 2002*		*Council census update 2018*	
	Location	*Year of construction*	*Shelters (n)*	*Total surface area in m²*	*Restoration areas*	*Number of family units recorded*	*Number of family units recorded*	*Number of residents*
Torrente Giostra/ Annunziata (north)	Villaggio Annunziata	1928	87	12,040.42	Area A: *Torrente* Annunziata	80	35	117
	Fondo De Pasquale	1925	37	3519.00	Area B: Giostra Ritiro *Tremonti*	765	668	2205
	Fondo Basile	1936	170	15,296.46				
	Ritiro	1938	222	63,495.76				
	Ritiro (Fondo Lauritano)	1927	215	12,729.00				
Torrente Zaera (Center)	Camaro S.Paolo	1927	188	19,928.18	Area C: Camaro Bisconte	991	739	2601
	Casalotto S.Luigi	1927	256	25,416.00				
	Bisconte	1937	298	88,624.15				
	Casalotto	1937	114	31,976.11				
	Fondo Tornatola	1927	46	6474.38				
Zona Sud—Maregrosso industrial zone	Fondo Saccà	1936	101	8536.91	Area D: Fondo Saccà	239	236	917

(*continued*)

Table 4.2 (continued)

Location of the torrent areas	*"Casette ultrapopolari" according to the 1960 legislative proposal*				*Council census 2002*		*Council census update 2018*	
	Location	*Year of construction*	*Shelters (n)*	*Total surface area in m^2*	*Restoration areas*	*Number of family units recorded*	*Number of family units recorded*	*Number of residents*
Torrente Gazzi (south)	Gazzi (Fondo Fucile)	1927	89	10,609.37	Area E: Gazzi Fucile via Taormina	1006	939	3585
	Gazzi (Rione Taormina and other areas)	1937	260	54,817.18				
Torrente san Filippo					Area F: Santa Lucia S.Filippo	50	29	90
Various areas					Area G: Bordonaro	139	76	229
	Total		2083			3336	2722	9744

Source: Our re-elaboration of different sources

Data on 1960: Legislative proposal De Pasquale, Li Cuasi, Pezzino, Gatto, Vincenzo, and Failla, Measures for restoration and slum clearance in the city of Messina, 6-4-1960, Chamber of Deputies

Data on 2002: Messina Council—local census: Territorial Policies Department, Priority order for interventions and chronological program for the Restoration of run-down areas in the City of Messina, Messina Council under Regional Law n.4 and 15-5-02, to amend and integrate Regional law n.10 of the 6-7-90

Data on 2018: Update Messina Council Census: Messina Town hall, Department for housing Policies. Mayor's report on the state of rehabilitation, 23/08/2018

Certainly, as D'Amico (2017) has observed, for the most part, today's shacks are not those of the earthquake, which in certain areas had been dismantled twice, in the 1950s and 1970s, and twice reconstructed. Nevertheless, it would sound a bit formalistic to neglect the evidence that the hundreds of *casette ultrapopolari* that were not torn down over the decades, as well as the shacks that had been rebuilt in the same, exact places by the new lumpen of the late twentieth century and, since then, handed down through generations, reflect a peculiar continuity. This continuity does not mean that today's 9744 residents are necessarily the descendants of the 15,000 from the past, although some of them probably are, nor that the shacks have remained unchanged since Ginatempo (1976) called them "doghouses". However, it is a continuity, or rather a persistence, that suggests that these "houses" and shacks were the tangible effect of formal and informal public policies, combined with private business interests, which have relegated the poorer classes to dormitory neighborhoods at the margins of the city (Cremaschi 1990).

In this perspective, the Messina catastrophe, as almost any other event of this sort, is interesting because, as we have already noticed, it helps make visible the ideology, the desires, and the aspirations of governments and planners. Reconstruction processes are, in fact, "rational" responses to large-scale solicitations that reflect specific values and visions upon desirable forms of order. Accordingly, as we have suggested in the Chap. 2, what makes disasters the perfect "ethnomethodological situation" is the tangibility of the institutional responses. If during "normal times"—the time during which crises are not so apparent—rulers and society produce changes bit by bit, in an almost invisible way and, sometimes, over the course of decades, in "post-disaster times" change is often produced with unstoppable fury. In modern times, however, such impetus is also said to be enclosed within frameworks of alleged rationality. For this reason, beyond being a tell-tale of the economic interests and the networks active in a given society, post-disaster management is also a "sub-cultural phenomenon" with political implications. A phenomenon, in other words, that pertains to specific segments of society (the "elites") and that sheds light on the common sense present within the most powerful societal circles. Post-disaster management, thus, is a cultural expression that can be read as a text, and it can also be understood as a radiant process that spreads cultures, common sense, and interests in the lower spheres of society.

But post-disaster management is also a collection of projects, or, rather, projections. That is, something that throws forth, in the time being, intents, desires, and effects. The disposition of bodies and classes in different spatial regimes, and the creation of two parallel cities, is, in the case of Messina, part of these projections and desires by the Fascist regime that will generate long-lasting and contradictory effects. Such an organization, in fact, reflected a double aspiration, or a double movement, aimed at producing a national middle class (or, perhaps, a petite bourgeoisie) and at preventing the revolt of those who were at the margins of local communities (Salvati 1994; Villani 2012; Tonelli 2013). The social structure that emerged by means of this project was partially new, for the introduction of a median element represented by this middle class (or, again, petite bourgeoisie), and the governmental "devices" utilized to include marginal and semi-marginal classes in the new social structure imagined by Fascism—these latter consisting of a peculiar and deranged use of welfare as a charitable and arbitrary technology, and a means for exchange (Villani 2012).

4.5 A Speculative Capitalism Based on Construction (1950–1990)

At the end of the Second World War, the supposedly temporary areas of shacks and shanties were continuing to expand. The chronic lack of housing, exacerbated by heavy bombing, had left a large number of people homeless. In addition, there was growing demographic pressure from the sheer numbers of peasants moving to the city because of the crisis in agriculture (see Chap. 2). According to the Istat Census of the population, in 1951, there were 222,166 residents, over 40,000 more than in 1931. Over the next decade, the population grew by a further 37,122 residents, to reach 257,288 inhabitants.

This demographic boom was reflected in the construction industry, also, in part, thanks to the extraordinary intervention for the South and international funds for reconstruction, which gave new vitality to the construction of subsidized public housing, as it was easier to manage spending flows. Public and private building rapidly came together, becoming a compact social bloc within a model that saw building as a means for creating business speculation and high profits and, at the same time, opportunities for political clientelism.

During this period, in Messina, a particular combination of the Keynesian-Fordist regulation model put the driving force of public spending and the State at the center of the economy, while in society, it was the stably employed male breadwinner working in a protected sector (public or large-scale industry) (Mingione 1991). As we know, the "Mediterranean" version of this model was based on certain specific characteristics (see, among others, Leibfried 1992; Saraceno 1994; Ferrera 1996, 2010; Castles and Ferrera 1996; Bonoli 1997; Andreotti et al. 2001), including:

- the low level of available services meant that families had to carry out many tasks themselves, receiving monetary compensation in return;
- the emergence of a particularistic and category-specific welfare system, in which social protection depended on the recognition of specific prerequisites and of belonging to certain social categories, that were often defined *ad hoc* on the basis of electoral cycles and the possibility of obtaining political support, trading benefits for votes, in a classic clientelistic exchange system;
- the importance of home ownership as a means to becoming a member of the middle classes and as a form of insurance for the future, especially in contexts such as those in the South, where lack of employment meant job instability and, as a consequence, only the minimum pension.

In the case of Messina, both public and private action united around the local construction industry. Just like the rest of southern Italy during the post-war period, and far from being an exception, Messina found "in building, in the process of producing urban residential dwellings and public works, a double specialization that was capable of supporting itself" (Bellicini 1990, p. 25). In particular, during the 1980s, southern cities became "residential cities, cities full of residential buildings, cities of houses" (ibid.), as evidenced by the imbalance between the quantity of non-residential and residential building, in favor of the latter. The southern economy was split between private construction, dependent on "residential mono-functionality" (ibid., p. 42), and public funding which, from the 1970s, was increasingly devoid of strategies, and geared toward supporting employment in an unproductive public tertiary sector or in the creation of public works, with the chaotic aim of never completing them, in order to maintain indefinite funding (Becchi 1990, p. 232). For example, in the second half of the 1980s, public works in the South made up

35–37% of the total of national works (a figure in line with the percentage of the resident population), but this figure went up to over 60% when it included the financial resources provided by the State (mostly for work relative to the extraordinary intervention) (Bellicini 1990, pp. 42–44). As Becchi (1990, p. 232) points out, by analyzing contract work for the two years 1989 and 1990, "60% of the total [...] was divided between just 24 provinces, 18 of which are situated in the Mezzogiorno with, [...] Messina overtaking Turin (1330 billion in comparison to 1260)". The most important aspect of this type of spending was to get the work started in order to stimulate the flow of expenditure, but not to complete it. As Becchi underlines (1990, p. 232): "the significance of the work therefore, [...] is not its utility, but the financial relationships, and not only, that can be created". Becchi continues that from this point of view, it is immaterial whether the work is finished, as long as it starts. In fact, it is enough just to begin the project, "that the work is announced".

This type of development model creates certain real effects within a territory, more particularly, an increase in uncompleted building projects that appear to have been "suspended" (Bellicini 1990, p. 44).

In absolute complementarity, the public and private actors in Messina sought to maximize their interests in terms of political consensus and financial gain. Clientelistic exchange could be introduced at any level and involve different social classes and might include the guarantee of a job in the building industry or help secure public housing, in fact, any activity, good, service, or acknowledgment that could be used to help stabilize the subject's social standing. Without forgetting that entrepreneurs or impromptu speculators could acquire contracts, speed up concessions, or receive help with bureaucratic practices.

The most interesting aspect was the precise division of the work, which made it possible to maximize profits. The best and most profitable areas of the city, that were more central and panoramic, were reserved for private construction, while the more marginal and peripheral areas (in the river valleys and southern side) were destined for public building projects and the creation of low-quality public housing through a system of contracts with private companies. Thus, the private companies were able to diversify the market and also to use public building contracts as a safety net for periods of decline in the private housing market, as Bortolotti had already noted in other cases (1978, p. 254).

In this context, public housing was built in the rural areas of the hinterland that had already been identified as the marginal areas suitable for the location of the post-earthquake shacks, the *casette ultrapopolari* and, finally, public housing. The areas involved were mainly the South, from the *Torrente* Gazzi up to and including the *Torrente* San Filippo; the center, above the *Torrente* Zaera, on both sides of the riverbed; and finally, in the northern part of the city, situated close to the *Torrenti Giostra* and *Annunziata*.

From the 1940s, and especially during the 1960s, large areas of public housing were erected, merging with the areas destined for rehabilitation and the local *fondi* of shacks and *casette ultrapopolari*. These were popular villages (Camaro, CEP village, Aldisio village, Santa Lucia), containing between 500 and 1000 dwellings. The southern part of the *Torrente* San Filippo was particularly affected by the presence of numerous blocks:

- the C.E.P. village (*Centro Edilizia Popolare*, Public Housing Center, a housing complex of about forty blocks built in a rural area once used for livestock breeding and farming);
- the *United Nations Relief and Rehabilitation Administration*, (UNRRA village; with reference to the international funds used for the construction);
- the *Istituto Autonomo Case Popolari* (IACP; Independent Public Housing Institute) complex in Santa Lucia, comprising over sixty low-income apartment blocks situated along the interior valley of the San Filippo torrent, in a large area next to the ancient rural village of Santa Lucia, and extending almost to the sea. The *Case Arcobaleno* (Rainbow Houses) were built in the 1990s and destined mainly for those living in shacks or houses in the San Filippo torrent, in an area that was cleared to make way for the construction of the new football stadium; finally, in 2010, a public housing block with sixty flats was completed in the lower part of the *Torrente*, which was partially populated after the clearance of the shacks in Fondo Saccà (see Table 4.2).

Without a development plan for these peripheral areas (it was only in 1978 that a new plan was approved, substituting the Borzì plan), this part of the city grew in a chaotic and disorganized way, outside the official city. This was the result of the successive "additions", the new buildings randomly erected next to existing shacks and self-builds, without any specific development plans (Campione 1988, p. 4), and guided exclusively by the opportunity to create income through public or private building contracts.

Their variability was influenced by the cycles of public spending and/or electoral rounds that either increased or decreased the amount of money available for funding from the public authorities in charge of constructing public housing, in particular, the IACP.

Speculative interests and the desire to maximize profits were the reasons for the unscrupulous cost containment that affected public and social housing construction from the 1950s onward, and which produced buildings that were increasingly inferior in quality, constructed using poor-quality materials, and rapidly deteriorated (Di Leo 1985). In the same way, there was a general lack of services, green spaces, and urban development, which turned these new areas into dormitory neighborhoods. Those assigned to public housing therefore found themselves relegated to shoddy houses in poor neighborhoods, with no services or opportunities. The post-war urban debate on the merits of planning public housing along the lines of Le Corbusier's utopistic model of *Unités d'intervention minimales* had envisaged the construction of housing that was able to guarantee social integration through the creation of social, cultural, and commercial services in the new popular neighborhoods. However, as often happens in other Italian and non-Italian cities, public housing projects are created without primary urban infrastructure (connections to the city center) or secondary (commercial, social, cultural, and recreational services, such as green areas with facilities, areas for sport, etc.) and with poor materials, accentuating the sense of segregation and social and spatial stigmatization for the people living there.

Figure 4.4 shows an aerial map of the city of Messina, where we outlined in black the popular districts that developed around the streams (the *Torrenti)*, which, from the mountains, go to the sea, cutting the city. In these districts the old shanties (the *casette ultrapopolari*) that resulted from the *fondi* policies mix with new self-built barracks, single low-income housing, and old and new social housing (IACP houses, CEP, or UNRRA houses).

Thus, Messina also possessed the typical features outlined by Bellicini with regard to the peripheral neighborhoods of the South in the 1980s (1989, p. 103 passim):

- its role as a residential reservoir, without the tertiary and commercial functions of the central and northern city suburbs;
- the prevalence of the "eternally unfinished", as if the "whole building were in the process of being built, erected temporarily, provisionally, that it is unfinished, and this not only applies to the legal,

Fig. 4.4 Aerial map of the city of Messina. Source: Our elaboration of ©OpenStreetMap contributors, licensed under the Creative Commons Attribution-ShareAlike 2.0 license (CC BY-SA) (30-01-2019). OpenStreetMap.org. Map Legend: The black lines outline the areas where the main popular districts are situated and include a mixture of IACP houses, shanties, and lower income housing

consolidated city, where there is no maintenance and no care for the exterior, but even more so to illegal developments, which constitute by far the larger part, and where the characteristics are even more exaggerated" (ibid., p. 107);

- the absence of services (which accentuates the residents' sense of precariousness) and the simultaneous presence of public housing and illegal building, which create a sort of perverse complementarity. Examples of this are "delayed or prolonged planning and construction of public housing which, in the meantime, enables illegal building to spring up in the same areas; the delay in refinishing the accommodation with all the necessary services, causing further delays in allocation, and, with the passing of time, making illegal occupation possible; and finally, a halt to the 'planned' implementation of basic services" (ibid., p. 104).

Speculative interests are also present in the complementary system that links private contractors to public housing construction through the system of contract tendering. This system strengthens the mechanism of reciprocal advantages between private companies and public developers.

Public administrations generally assign work they are unable to carry out themselves through public contracts, so the contract is put out to public tender, specifying the requirements to be met by all participants, along with certain guarantees in case they win the contract, and a price limit that must not be exceeded. The contract procurement system has often been analyzed as an expression of systematic corruption in Italy (Centonze 2005; Della Porta and Vannucci 2007). Different mechanisms are often used to rig or direct the tenders in favor of certain participants, at the expense of others, and in exchange for the payment of bribes or political favors. The public or political actor responsible for guaranteeing the successful outcome of the tender is set to gain huge benefits; these may be monetary and/or boost their political consensus or personal prestige within clientelistic networks (which may prove to be useful in the future). Private entrepreneurs are thus able to operate in protected markets, against fictitious competition that does nothing to stimulate them to improve. In fact, winning the contract has nothing to do with the quality of the service on offer; it is dependent on the ability to successfully gain access to a clientelistic network (Becchi and Rey 1994; Centorrino 1995). A transaction of this type has a high cost that is usually added to the price of the work to be carried out, which is already of very low budget because of the way the contract system works. As mentioned earlier, in the case of public housing projects in Messina, this often means that construction materials are substandard and tend to deteriorate within a few years, or even during construction. In addition, some of the complementary building works that would greatly improve the quality of the residents' lives are never carried out so that savings can be made.

Up to now, the focus has mostly been on public building, but it is also important to underline that the private companies always secured the most expensive building land, especially the areas in the city center. At the end of the 1950s, the saturation of the Borzì master plan brought about a change in private property strategies, which aimed to change the type of action while still guaranteeing full profit maximization (Di Leo 1985). The approval of several changes to national anti-seismic legislation allowed some of the restrictions contained in the Borzì plan to be relaxed; more specifically, the maximum building height was raised (as it had been during the 1930s), to a height of eighteen meters above ground level. Following these changes to the law, buildings were constructed in the few empty areas that remained and were juxtaposed with older buildings without any coherent plan or sense of continuity. Worse still, the older, one-

and two-story buildings dating from the 1800s were demolished, without any consideration for their artistic or cultural value, to make way for new five-story blocks.

Moreover, in the search for new building space, private housing for the middle and upper-middle classes increasingly compromised the scenic hills behind the city, a location that was also exploited for the construction of a new ring road, to the North of the city, the "Panoramica dello Stretto", so called because of the breathtaking views across the straits of Messina. This area became the new construction zone for luxury houses, and, from the hills, they seem to dominate the city, an expression of how the city has been despoiled.

It is also interesting to note that these better-quality residential areas are often built without primary or secondary urban facilities and infrastructures. Green spaces, parking, and other services are designed as external components in private housing complexes, and they are considered private benefits for the residents rather than collective assets open to all. There is also very limited public transport connections with the rest of the city. The speculative side of building is clear here, in this privatistic concept of space, where the important thing is to maximize the value of individual property rather than collective space, but it is also an expression of the local building speculation mentioned earlier.

The speculative side of the building industry is also visible in the large number of bankruptcies among building companies. From the late 1960s, many private construction companies sold on paper only, and after taking advance payments for the property from the buyers, would declare bankruptcy, leaving the work unfinished. This tried and tested system allows contractors to collect sums of money that will never be paid back (because they are tied up in starting the work). Often, contractors close down one business and open another with nominee holders, continuing to operate on the local market even when they have been banned. In turn, this creates opportunities for contractors to buy back the unfinished buildings at auction for a very low price so they can finish them.

A further sign of building speculation can be seen in the gradual relaxing of restrictions, in the special exemptions and lack of official controls, which lead to serious building irregularities and the construction of illegal buildings, and they are often resolved through regular building amnesties. Suffice it to say, that it has become common practice for private owners to enlarge their apartments using various stratagems, such as illegal extensions and attics built on terraces, which start out as removable structures and then become permanent, or mezzanines without the necessary authorization. There is also a widespread trend in building illegal second homes by the sea (Gucciardo 1996, p. 252). Far from being penalized, these illegal actions are regularly pardoned through building amnesties. Unfortunately, building amnesties are used so often in Italy to create revenue and resolve numerous irregularities that they end up being an incentive for opportunistic behavior, which is particularly cost-effective (Donolo 2001).

This point deserves a little expansion. Illegal building practices are extremely widespread in Italy. As Cremaschi underlined (1990, pp. 141–142), using the Cresme (1988) survey carried out by the Ministry of Public Works, to evaluate the effects of Law 47/1985 on the amnesty of illegal building work, during the 1950s about 65,000 illegal houses were built. This figure doubled during the subsequent decade, so that for the period 1950–1983 (the reference period used), about 14% of the overall number of dwellings was illegal (3,300,000 dwellings). Added to this were the seven million illegal transformations of existing buildings (changes to the intended use, restructuring, small modifications). In fact, in Italy, practically one family in three had found it necessary to bend the construction rules. In reality, as Gambuzza (1986, p. 67) has observed, the term "illegal building" is a cultural and technical concept of urban planning that refers to legal principles not dissimilar to those found in common criminal discourse, implying a vision centered on irregularity. *Spontaneismo* (or spontaneous construction, that is, building without planning permission, following one's needs and aspirations), on the contrary, appears to be an expression more suited to describing a city environment in which it is difficult to find a shared meaning of the law and a public figure who clearly embodies the notions of legitimacy and order rather than legality itself. However, it might be a good idea to distinguish now and again between illegal building created through "need" and "speculative" "illegal building".

From the 1970s, and especially in the South of Italy, the phenomenon of *spontaneismo* assumed epic proportions: in the period 1971–1981, Cresme (1988) estimated that a good 61.6% of dwellings built in the South were illegal, or rather, had no planning permission. Gucciardo (1996, p. 250) calculated an estimate for the subsequent decade (1981–1991) and found there was still a very high percentage of illegal buildings (51.7%), despite the 1985 amnesty. More specifically, Sicily was the region with the highest rate of illegal building: between 1971 and 1984, with 27.3% of illegal housing (ibid., p. 249).

During those years, a series of economic circumstances (inflation, a rise in interest rates, the rent regulation law, which lowered rent prices to the detriment of property owners, thus drastically reducing the rental market), as well as the State's seemingly high tolerance to illegal building (which was rarely penalized), made house purchase uneconomical and, in a certain sense, gave rise to illegal self-building (Gucciardo 1996, pp. 251–253). *Spontaneismo*, thus, became a phenomenon that cut across all social classes. For example, Fera and Ginatempo's (1985, p. 275) research on Sicily showed the transversal nature of illegal self-building. In their research, they noted that only 24.4% of the self-builders were precarious workers or unemployed, while 34.8% were salaried workers, 26.4% were self-employed, 8.2% were civil servants, and 6.3% business owners, freelance workers, officials, and managers. Illegal self-building allowed savings of up to 40–50% (ibid., p. 51), which is why it was chosen as a strategy, even by those who were not in a condition of need. As far as Messina is concerned, we may well ask what the difference is between an illegal shack built by a low-income family belonging to the lower classes, in a popular peripheral neighborhood, and an illegal extension built by a middle-class or even upper-middle class family, a practice that is widespread in the city center. At first sight, it would seem very little, but only the former is considered an expression of incivility and lack of collectivity, while, on the contrary, the latter is widely tolerated, almost to the point of not considering it a crime. This difference in treatment reflects the high social acceptability of so-called white-collar crimes (among others, Sutherland 1949; Lascoumes and Nagels 2014) and also reflects the classist attitude that condones the illegal building of an extension on the veranda of a middle-class apartment (subsequently given amnesty), because it is probably simply an improvement to the property. However, if the same thing happens in a popular neighborhood, for most commentators, then it most certainly expresses deterioration and particularistic appropriation.

4.6 Continuity, the Neoliberalization of Building

As summarized in Chap. 3, the Fordist regulation model entered a period of crisis in the 1990s. The emergence of financial globalization was accompanied by the need for financial rigor in order to control public debt and remain within the strict parameters of Maastricht, in view of the European Union (EU) monetary agreement, and to avoid speculation on government securities in the global market. The discovery of systemic corruption in Italian politics and the public administration with the Clean Hands scandal showed there was a need to reform the political sector. This was especially so in the South, where public government jobs and public spending had supported families by means of clientelistic exchanges geared toward the reproduction of political consensus and the stabilization of the social fabric through forms of particularistic inclusion.

In Messina, the crisis in the Fordist model could be seen in the growing inability of the public sector to act as a sponge, directly or indirectly supporting citizens' incomes and guaranteeing forms of inclusion and social mobility of a particularistic nature, even to the more external, marginal circles (see Chap. 4). The breakdown in the Fordist building model is visible in the block on public spending and the impossibility of setting up large-scale public sector projects because of lack of funds, generating a positive complementarity between private and public building.

The first move was to try to stretch the model by using alternative funds rather than national ones, but once the season for the building of large popular neighborhoods was over, the game re-opened with the "rehabilitation" of the shanty areas, with the Sicilian Regional Government at the forefront of the field. Regional Law n.10 from 1990 (Regione Sicilia 1990), regarding rehabilitation plans for Messina gave the Council the task of identifying the "environments" most in need of restoration (already generally defined by the Tekné master plan in 1978) and requested they put forward detailed plans for each area, with a budget of 500 million lire for slum clearance and requalification. It was not until 2002 that the city finally managed to identify the areas, carry out a census of the families living there, and get approval for the detailed plan. The delay in this bureaucratic process meant that spending opportunities were frustrated. Of the initial 500 million, only 150 were spent, to which a further €70,000 was added in 2004 by the region (Redazione 2013).

Moreover, between the late 1980s and the early 1990s, some lukewarm support for the building industry came from self-financing building cooperatives, which, despite their many limitations and ambivalences, seemed partially able to guarantee the sort of liquidity that had been missing since public expenditure had dried up and public building had slowed down.

However, it was only with the opportunities opened up by the new neoliberal and de-regulation policies that it was possible to re-establish the development model. In 2002, after a lengthy process lasting 10 years and 700 amendments, PRG, the new urban development plan, was adopted (Comune di Messina 2002), liberalizing the property market and permitting the launch of a speculative bubble with property construction at its heart.

The new urban development plan was decidedly inflated, estimating an increase in the population of up to 290,000 inhabitants despite the fact that both the Istat Censuses of 1991 and 2001 had already shown a drop in the number of residents. The predicted demographic trend was emphatically refuted by both general national statistics and statistics regarding the city itself. However, the prediction helped to justify what would otherwise have been an unfounded expansion in construction, which, as is seen, permitted excessive land consumption, also in hydrogeological high-risk areas. The new development plan further increased building levels, and also proposed creating zones within the city, whereby entire central and peripheral areas were defined as expansion or completion zones—including the areas in the hills or near the sea—thus freeing up many areas that had been considered "areas of natural beauty and interest" in previous plans, prohibiting building or, in some cases, excessive building. As a rule, in almost all parts of the city, it became possible to increase building height and the number of floors, which could reach as many as seven floors above ground level in some parts of the south, north, and even in the center. Moreover, parking lots and attics were not included in the calculation for the number of floors. Paradoxically, the floors reserved for parking and garages were not included in the building index established for each area, but it was possible to change the intended use. In effect, as happened in the southern part of the city, buildings that were designed with seven floors were, in reality, constructed with ten floors if we consider the basement floors (for cellars or garages), which incorporated apartments, and attic floors. These were huge buildings, which stood out in comparison to previous constructions. In over sixty pages, the development plan was principally dedicated to the zoning of the city and to stipulating rules (or rather, weakening existing ones) for the construction of apartment blocks, without suggesting any plans for the development of services, green spaces, sports facilities, or general enhancement of the territory (Comune di Messina 2002).

In fact, it was possible to build in any area, even in those classed as hydrogeological high-risk areas or areas of natural interest. For example, the area near the Straits of Messina (Capo Peloro), that was designated a "Nature Reserve", in 2001, and for which the plan established that "seasonal accommodation or hotels" could be built, opened the way for the division of entire portions of the hillsides in order to build expensive villas overlooking the sea, in an area in which the middle and upper-middle classes generally spent their summers (Caspanello 2018).

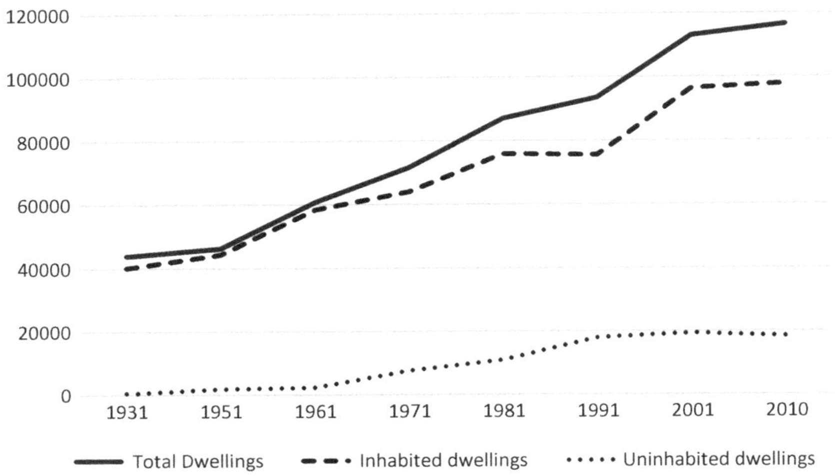

Fig. 4.5 Total number of dwellings, inhabited and uninhabited, 1931–2011. Source: Our re-elaboration of Istat Census Data (different years)

This neoliberalization of the building sector encouraged rampant residential construction, once again with little or no concern for the realization of primary and secondary infrastructures.

From the approval of the development plan to the 2011 Istat Census of the population, there was a rise in the number of houses of +4.16%, in comparison to −3.5 residents. Thus, the overall number of empty dwellings was increasing, reaching 18,429 in 2011, that is, 15% of the total. As shown in Fig. 4.5, from the scarcity of post-war housing (which implied that almost all the available houses were occupied), there was a gradual saturation of the market which started in the 1990s, with a rise in the number of empty dwellings.

The boom in private construction, triggered by the de-regulation of building concessions and sanctioned by the development plan, gave rise to a veritable speculation bubble, with a huge rise in the value of property. This only began to slow down in 2008, owing to the international economic crisis and following the 2009 hydrogeological disasters that hit two different parts of the Messina district (Giampilieri and Scaletta Zanclea). This highlighted the fragility of the territory caused by the impact of anthropic intervention and overbuilding, especially close to riverbeds and abandoned rural areas that have consequently become subject to landslides (Falconieri 2017).

National housing policies wanted to respond to the lack of housing and overcrowding that was rife in the 1950s by expanding the housing stock and promoting home ownership, by any means necessary, and through support for the sector and a chaotic system of benefits, exemptions, and privileges (Cremaschi 1990, p. 147). These implicitly supported the development model based on residential building that continued to be self-sustaining even when, at the beginning of the 1980s, mass immigration marked an irreversible demographic inversion. The effect of these policies on a city such as Messina meant that there was a huge increase in the number of property owners and a gradual decrease in the number of tenants (Fig. 4.6).

However, our idea is that in the case of Messina, it is not important to stress on novelties. Rather, the story of this city is, as often happens, a complex and coherent case of continuity and overlapping. As stated by Ginatempo (1976), post-earthquake urban planning was based on a clas-

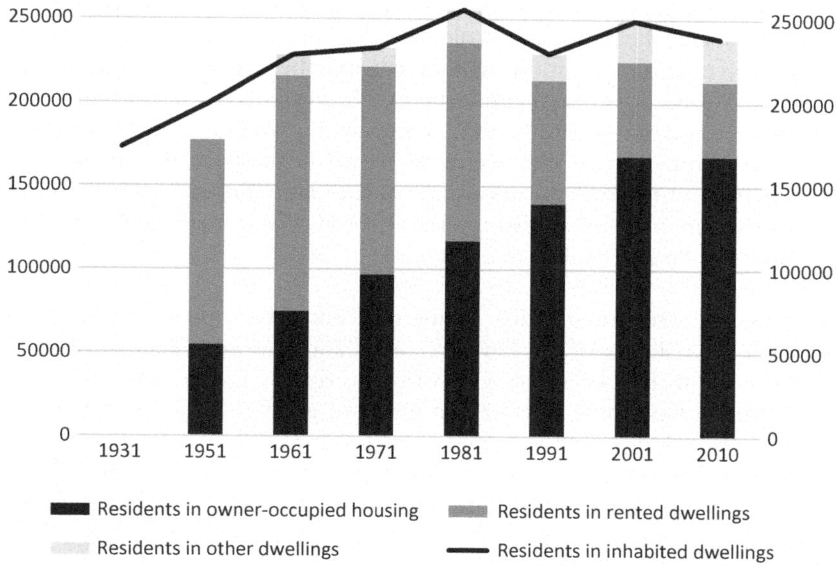

Fig. 4.6 Resident population according to accommodation in use (1931–2011). Source: Our re-elaboration of Istat Census Data (different years)

sist vision of the city, showing that there was a continuum in the city's housing policies, from the very first measures set out by Giolitti, to the Borzì plan, the Fascist "rehabilitation" policies, and the subsequent post-war plans. This vision can be summarized in two related concepts that are never openly declared but are implicit in the attitude toward the people still living in the shanties:

- The lower classes cannot aspire to living in the historical center of the city; they must be relegated to the margins. Surprisingly, this is a concept that connects the zones designated for popular housing in the 1960s with contemporary urban rehabilitation plans, which aspire to the gentrification of these areas, until the poorer classes are entirely removed from this urban fabric.
- After all, families who need public housing because their homes were destroyed by an earthquake, but who do not possess the economic means or a job to buy a house independently—the "needy", "poor" families—have to make do with temporary housing ("casette ultra-popolari"), which, in any case, corresponds to their social class, and might even represent an "improvement" to their social condition. In this sense, the urban sub-proletariat should think themselves lucky to receive accommodation, whatever it may be, especially because it is not always possible to verify their real, presumed, or artificially constructed condition of "need". Once again, we find an expression of the punitive logic that seems to lie at the basis of the distinction between the *deserving* and *underserving poor* (Polanyi 1944) which, as is seen in Chap. 7, distinguishes a lot of common discourse regarding the "right" to live in a shack.

Thus, the creation of this segmented and exclusionary space, that impedes the visibility of the "underclasses" and imposes a regime of controlled mobilization of an army of workers, voters, and beggars, is at the same time old and new. It is new in terms of geography and distribution of poverty, and also for the inclusion of the lower classes in the political arena—prior to the advent of Fascism, and as a consequence of the introduction of universal suffrage, and afterwards, as an aspect of the Fascist project for the reconstruction of the Italian nation. Nevertheless, over the course of the twentieth century, the substance of the relation between the State and the classes remains the same in terms of instrumentality. The masses had to be managed and mobilized in the best interest of the State's elite, or of the groups of interest that ruled the city. Thus, charitable measures were the carrots, while exclusion and repression being the stick. Torrente Giostra (Fondo Basile) (Di Fascist welfare, perhaps, is the most important "technology" introduced in the aftermath of the disaster. The tool that, in the passage to democracy, will be inherited by the Christian Democrat governments and used as an instrument to obtain consensus among the southern underclasses (Sassoon 1997, pp. 35–36). In this respect, Garcia and Karakatsanis (2006, pp. 99–100) wrote that the tactic of the Christian Democrats consisted in fragmenting the occupational social security system and so increasing particularism. Within this segmented framework, patron-client relationships operated through the various branches and agencies of public administration in order to allocate resources and benefits (labor, pensions, doles, etc.). Therefore, from the post-war era to very recent times, social security measures and institutions have been the instrument through which local constituencies of support were built up. A clear Fascist legacy that has influenced the Italian democracy for the time being (Garcìa and Karakatsanis 2006, p. 101; Paci 1984).

If these were, after all, general tendencies that apply to most of southern society, it is the local insertion of Messina within this moral economy which is peculiar, being the 1908 earthquake the event that triggered this dynamic (see Chap. 1)—coupled with the slow pace of the reconstruction, which had crossed different political eras and had, therefore, been intimately shaped by the different political phases that had characterized the country since the beginning of the seismic crisis.

However, the sense of continuity is provided also by the "stability" of the city economy. It is not by chance that, in 1985, Mario Centorrino, an eminent local economist, at the end of an article could ask whether Messina was "a reconstructed city, or a city to be still rebuilt?" (Centorrino 1985, p. 151). In the same article, Centorrino listed and analyzed the economic plans elaborated since the 1960s by the local and regional development agencies. The conclusion of the author was that there has never been a plan. In fact, not only none of the previous plans were actively pursued, but each of the documents prepared by different agencies contradicted the others. For example, if the regional government imagined a development based on the tertiary sector, the local plans referred to the "industry". An "industry", moreover, that was never located in the city, but in the province, and that contemplated Messina as the space of the new services—all this, within a scenario that lacked fundamental infrastructures. As Sgroi (1978, pp. 91–92) put it, in this very scenario, the few large investments by private investors were not coordinated and had often a speculative nature, mostly based on the use of public incentives. These were investments, moreover, which were concentrated for the most part in the province and whose frequent outcome was the default of the firms and the use of state-funded temporary lay-off schemes for employees. In this situation—added Sgroi (1978, p. 113)—even the weaker components of the middle class were marginalized, and they displayed the features of a "lumpen-bourgeoisie". In this regard, anticipating analyses and debates that became popular only during the 2000s, the same scholar referred to the crisis of the knowledge workers in the stagnant economy of the city, and to the separate effects that the different availability of weak and strong ties had on the younger members of the lower segments of the local middle class—a combination of elements that had emigration as a natural outlet. A movement of population confirmed by the 1971 census (Chiara 2011, pp. 171–172), the first to show the relevance of a phenomenon that will reach its peak over the current decade—when, since 2011, the population will lose between 1000 and 2000 people every year, shifting from 247,000 to 236,000 inhabitants in 7 years. Figures, however, that do not include those who have moved without having changed their residency (Redazione 2018).

However, continuity is also provided by the persistence of the shacks and "casette ultrapopolari" in the city landscape. In 2002, the Council Census for the application of the rehabilitation law identified 3336 family units and around 11,500 residents.[12] Sixteen years later, there are still 2722 family units and 9744 residents (Comune di Messina 2018). From 1990 up to the present, after almost 30 years of important public investment in public housing, only about 800 public dwellings have been handed over.

These dilapidated houses are the direct consequence of that territorial segmentation that started in the 1910s, was consolidated by Fascism, and was followed by the building boom of the post-war years. In short, those years saw the rise of the first of these peripheral units, in a geographical and/or social sense, which still exist and serve as a backdrop to the lives of many of the protagonists of this study. In local linguistic codes, these housing units and the neighborhoods within which they develop have become partly or completely synonymous with marginality, crime, poverty, and, later, even with the concrete expansion, that is, vast agglomerations of mostly gray and decaying public housing, alternated to shacks and housing of a speculative "residential" type.

As we have underlined throughout this chapter, these neighborhoods are part of a general development model that creates profit through the consumption of the territory, building speculation, and property income, a model that is able to move between public spending and neoliberalism.

In addition, it is precisely within this model that we can discern certain elements of continuity that are present in the peripheral areas and also, surprisingly, in the city center and in the new middle-class residential areas. More specifically, these are the total absence of non-residential building and the general lack of infrastructure and services; the devastation of the hillsides behind the city (with the consequent increase in the risk of hydro-geological disasters); the juxtaposition of extraordinarily different building styles, buildings that are worlds apart and do not belong in the same urban environment; the growing weight of illegal building (which may well be condoned, but it is disorderly and chaotic, thanks to the addition of further elevations to apartment buildings, the enclosure of verandahs and enlargement of balconies, etc.).

Behind this city of houses, of apartment blocks that deny people the opportunity to enjoy shared social spaces, it is possible to see how the buildings express the function of social regulation, which is quite different to what one might expect on paper (i.e., the population's need for housing). Behind the rhetoric about the development of the city, about improvement to the housing stock, and about "rehabilitation", construction performed a series of important social regulation functions, to help integrate local society. It acted to reinforce the social bloc between the private businesses and public institutions that governed the city, guaranteeing to both financial income and power, through a network of reciprocal exchanges and favors. It underlined the stratification of the city from a spatial and geographic point of view, drawing a line between the poorest citizens, who were relegated to the ghettos of public housing on the margins of the city, and the upper-middle class, who resided in the elegant houses in the city center or the private residential buildings situated in the hills. Finally, it was one of the means for guaranteeing forms of particularistic inclusion for the working class—a precarious job as a laborer, the possibility of access to the public housing waiting list or of being on top of the list and receiving a house, all represented the terms for clientelistic exchange which, however unfair or unbalanced, represented a real opportunity for the client's social mobility. "Give us a job, or a house, or both" was the perimeter for the only possible claim (individualistic or privatistic) in an exchange system where the payoff was not simply one individual vote, but the votes of a whole network of family and/or friends.

In the next chapter, we will deal with the subsistence strategies put in place by the subjects living in these lower-class areas and who work as flexible or precarious service workers or laborers.

Notes

1. Heidegger (2008, p. 62), De Martino (2015, p. 85 passim).
2. As anticipated in Chap. 3, the industrial processing of reject lemons was very widespread in Messina, also linked to the production of citrate (Lupo 1987, pp. 84, 93).
3. As Harvey sums up (2010, p. 29): "As more surplus capital went into production during the 1980s, particularly in China, heightened competition between producers started to put downward pressure on prices (as seen in the Wal-Mart phenomenon of ever-lower prices for US consumers). Profits began to fall after 1990 or so in spite of an abundance of low-wage labor. Low wages and low profits are a peculiar combination. As a result, more and more money went into speculation on asset values because that was where the profits were to be had. Why invest in low-profit production when you can borrow in Japan at a zero rate of interest and invest in London at 7 per cent while hedging your bets on a possible deleterious shift in the yen–sterling exchange rate?".
4. In this text, the uses of world "barrack" and "shack" are subjected to the local semantics. By barracks we mean the technical artifact; that is, a temporary wooden structure provided by the government in the aftermath of the earthquake, aimed at hosting households and individuals. By the term "shack" (or shanty, and similar) we mean the process of permanentization and degradation of the artifact and the social space.
5. For a technical analysis of the master plan, see: Licordari (2017).
6. This torrent divided into two at a certain point, and it had two beds (Giostra and Annunziata).
7. Initially, the areas destined to private housing development were Corso Cavour, Porta Imperiale West, Santa Cecilia South, Primo Settembre North, and La Farina East. As soon as these areas were filled, the space for the bourgeoisie expanded: north of Via Primo Settembre, east of Corso Vittorio Emanuele, north of *Torrente* Giostra, west of the beltway, and

around the most important public buildings (courthouse, city hall, government houses etc.).

8. As soon as the areas were filled, the expansion of this type of houses included: Viale Europa, Piazza del Popolo, Piazza Duomo, Via Romagnosi, Via Oratorio della Pace (Boccetta South) and Piazza San Vincenzo (Boccetta North).
9. The areas selected in the original plan were between Villa Dante and Gazzi, in the southern section of the city. Later, they were also built along Via Tommaso Cannizzaro, the beltway, Boccetta, and Via Garibaldi.
10. Originally, these houses could be found in the surroundings of Villa Dante and along the Giostra and the Trapani torrents. Later, other series of buildings would be erected near the beltway, along the Giostra Torrent (Fondo De Pasquale).
11. Initially, these houses could be found in *Torrente* Gazzi, Fondo Vadalà, Fondo Tornatola, Fondo Lauritano, Fondi Martinez-Consiglio, Villaggio Matteotti (near *Torrente* Annunziata-Giostra). Later, they invaded the other underclass spaces, and were also built near Villa Dante, and all along Via La Farina as far as Viale Europa (Torrente Zaera), and *Torrente* Portalegni. Also, to the north, at the top and the bottom of the beltway, along the Giostra torrent (Fondo de Pasquale). And, in the late 1930s, also in other parts of Gazzi (Mangialupi, Via Taormina), *Torrente* Zaera (Bisconte, Ritiro) and *Torrente* Giostra (Fondo Basile) (Di Leo and La Spada 1985).
12. We have not used the data from the latest Istat Census since it is not very reliable. The 1971 Census revealed that there were 3344 "unfit dwellings". In 1981, the number decreased to 635 and a few units (27 in 1990 and 72 in 2001) in the following decades, only to increase to 1170 in 2011, but a much lower number than the data from the Messina Municipality shows as far as rehabilitation is concerned. Such fluctuating data that underestimates the number of shacks is probably due to errors in the survey, so that many of the *casette ultrapopolari*, which fall into the rehabilitation areas were obviously registered in the census either by the officials or by the residents, as housing.

References

Adler, P. A., Adler, P., & Fontana, A. (1987). Everyday Life Sociology. *Annual Review of Sociology, 13*, 217–235.

Andreotti, A., Garcia, S. M., Gomez, A., Hespanha, P., Kazepov, Y., & Mingione, E. (2001). Does a Southern European Model Exist? *Journal of European Area Studies, 9*(1), 43–62.

Baglio, A. (2005). *Il Partito Nazionale Fascista in Sicilia. Politica, organizzazione di massa e mito totalitario 1921–1943.* Manduria, Bari, and Rome: Lacaita.

Barbera Cardillo, G. (1981). *Messina dall'Unità all'alba del Novecento.* Genève: Librairie Droz.

Barone, G. (1982). Sull'uso capitalistico del terremoto: blocco urbano e ricostruzione edilizia a Messina durante il Fascismo. *Storia Urbana, 19,* 47–104.

Battaglia, R. (2003). *L'ultimo "splendore". Messina tra rilancio e decadenza.* Soveria Mannelli: Rubbettino.

Becchi, A. (1990). Opere pubbliche. *Meridiana, 9,* 223–243.

Becchi, A., & Rey, G. (1994). *L'economia criminale.* Rome and Bari: Laterza.

Bellicini, L. (1989). In periferia. Temi, percorsi e immagini. *Meridiana, 5,* 93–128.

Bellicini, L. (1990). La costruzione del territorio meridionale. *Meridiana, 10,* 11–44.

Bevilacqua, P. (1981). *Catastrofi, continuità, rotture nella storia del mezzogiorno.* Torino: Einaudi.

Bonoli, G. (1997). Classifying Welfare State: A Two-Dimension Approach. *Journal of Social Policy, 26*(3), 351–372.

Bortolotti, L. (1978). *Storia della politica edilizia in Italia.* Rome: Editori Riuniti.

Braudel, F. (1984). *Civilization and Capitalism, 15th-18th Century, III: The Perspective of the World.* New York: Harper & Row.

Brighenti, A. M. (2010). *Visibility in Social Theory and Social Research.* New York and London: Palgrave Macmillan.

Budà, D. (2012, October 3). Messina e le sue fiumare, evoluzione storica e prospettive (Parte III). *Tempostretto.* Retrieved from https://www.tempostretto.it/news/speciale-tempostretto-messina-sue-fiumare-evoluzione-storica-prospettive-parte-iii.html.

Caldo, C., & Santalucia, F. (1997). *La città meridionale.* Firenze: La Nuova Italia.

Campione, G. (1988). *Il progetto urbano di Messina: documenti per l'identità 1860–1988.* Rome: Gangemi.

Caspanello, A. (2018, April 20). Messina, come il piano regolatore ha deturpato il volto della città. *LetteraEmme.* Retrieved from www.letteraemme.it.

Castles, F. G., & Ferrera, M. (1996). Home Ownership and the Welfare State: Is Southern Europe Different? *South European Society and Politics, 1*(2), 163–185.

Centonze, A. (2005). *Il sistema di condizionamento mafioso degli appalti pubblici. Modelli di analisi e strumenti di contrasto.* Milano: Giuffré.

Centorrino, M. (1985). Messina, economia e sviluppo degli anni '60. In G. Di Leo & C. M. Lo (Eds.), *Messina, una città ricostruita. Materiali per lo studio di una realtà urbana* (pp. 147–151). Edizioni Dedalo: Bari.

Centorrino, M. (1995). *Economia assistita da mafia.* Soveria Mannelli: Rubbettino.

Checco, A. (1990). L'età della decadenza (1861–1945). In P. Orteca (Ed.), *Il porto di Messina dalle origini ai nostri giorni* (pp. 75–142). Messina: Edas.

Chesnais, F. (1996). Mondialisation du capital et régime d'accumulation à dominante financiére. *Agone, 16*, 15–42.

Chiara, L. (2011). *La modernizzazione senza sviluppo. Messina a cento anni dal terremoto*. Firenze: Le Lettere.

Cicciò, S. (2016). *Il porto di imbarco di Messina: L'ispettorato e i servizi di emigrazione*. Milano: FrancoAngeli.

Comune di Messina - Dipartimento di politica del territorio, Direttore del Dipartimento, Settore Urbanistica e gestione del territorio. (2002). *Piano regolatore generale. Norme tecniche di attuazione*. Approved with D.D.R. n.686/2002 e n.858/2003. Messina: Comune di Messina.

Comune di Messina - Dipartimento Politiche per la casa. (2018, August 23). *Relazione al Sindaco sullo stato del risanamento*. Messina: Comune di Messina.

Cremaschi, M. (1990). L'abusivismo meridionale: realtà e rappresentazione. *Meridiana, 9*, 125–154.

Cresme. (Ed.). (1988). *Indagini rivolte alla conoscenza degli effetti della legge di sanatoria dell'abusivismo*. Ministero dei Lavori Pubblici.

Cuaresma, J. C., Hlouskova, J., & Obersteiner, M. (2008). Natural Disasters as Creative Destruction? Evidence from Developing Countries. *Economic Inquiry, 46*(2), 214–226.

Cutter, S. L. (1995). Race, Class and Environmental Justice. *Progress in Human Geography, 19*(1), 111–122.

D'Amico, L. (2017, November 19). Fondo Fucile e quei "belvedere" del degrado. *Gazzetta del Sud*. Retrieved from http://www.gazzettadelsud.it/news/messina/267389/fondo-fucile-e-quei-belvedere-del-degrado.html.

D'Angelo, M. (1995). *Comunità straniere a Messina tra VIII e XIX secolo. Alle origini del British Cemetery*. Messina: Perna Edizioni.

D'Angelo, M. (2009). Prima e dopo. Messina 1902–1914. In G. Campione (Ed.), *La furia di Poseidon. Messina 1908 e dintorni*. Milano: Silvana Editoriale.

De Certeau, M. (1990). *L'invention du quotidien. 1 - Arts de faire*. Paris: Gallimard.

De Martino, E. (2015). *Magic. A Theory from the South*. Chicago: Haubooks.

De Pasquale, D. (1992). *Modalità e tecniche di accumulazione fondiaria a Messina nell'Ottocento*. Messina: Sikelia.

De Pasquale, D. (2006). *Le mani su Messina. Prima e dopo il terremoto del 1908*. Messina: Armando Siciliano Editore.

De Pasquale, Li Cuasi, Pezzino, Gatto, Vincenzo, and Failla. (1960, April 6). *Proposta di Legge "Provvedimenti per il risanamento e lo sbaraccamento della città di Messina"*. Camera dei Deputati.

Della Porta, D., & Vannucci, A. (2007). *Mani impunite. Vecchia e nuova corruzione in Italia*. Rome and Bari: Laterza.

Di Leo, L. G. (1985). Stratificazione urbana dal secondo dopoguerra e recupero dell'esistente. In G. Di Leo & M. L. Curzio (Eds.), *Messina, una città ricostru-*

ita. Materiali per lo studio di una realtà urbana (pp. 45–48). Bari: Edizioni Dedalo.

Di Leo, G. L., & La Spada, E. (1985). La città dopo il terremoto: interventi pubblici e stratificazioni del patrimonio edilizio. In G. Di Leo & M. L. Curzio (Eds.), *Messina, una città ricostruita. Materiali per lo studio di una realtà urbana* (pp. 11–18). Edizioni Dedalo: Bari.

Dickie, J. (2000). A Patriotic Disaster: The Messina-Reggio Calabria Earthquake of 1908. In G. Bedani & B. Haddock (Eds.), *The Politics of Italian National Identity. A Multidisciplinary Perspective* (pp. 50–71). Swansea: University of Wales Press.

Dickie, J. (2002). The Smell of Disaster: Scenes of Social Collapse in the Aftermath of the Messina – Reggio Calabria Earthquake, 1908. In J. Dickie, J. Foot, & F. Snowden (Eds.), *Disastro! Disasters in Italy since 1860. Culture, Politics, Society* (pp. 235–255). New York: Palgrave.

Dickie, J., & Sayer, D. (2005). *Messina: A City without Memory?* Retrieved from http://www.ucl.ac.uk/place-and-memory/messina/index.html.

Donolo, C. (2001). *Disordine*. Rome: Donzelli.

Dyl, J. L. (2017). *Seismic City. An Environmental History of San Francisco's 1906 Earthquake*. Seattle and London: University of Washington Press.

Falconieri, I. (2017). Smottamenti. In *Disastri, politiche pubbliche e cambiamento sociale in un comune siciliano*. Rome: Cisu.

Farinella, D. (2009). Messina, un'economia in transizione tra cambiamento e marginalità. In M. Centorrino & P. David (Eds.), *Le città della Fata Morgana. 5° rapporto sull'economia della provincia di Messina: analisi socio-economica e stili di vita dell'Area dello Stretto* (pp. 47–84). Milano: Franco Angeli.

Fera, G., & Ginatempo, N. (1985). *L'autocostruzione spontanea nel Mezzogiorno*. Milano: FrancoAngeli.

Ferrera, M. (1996). The 'Southern Model' of Welfare in Social Europe. *Journal of European Social Policy, 6*(1), 17–37.

Ferrera, M. (2010). The South European Countries. In F. G. Castel, S. Leibfried, J. Lewis, & C. Pierson (Eds.), *The Oxford Handbook of the Welfare State* (pp. 616–629). Oxford: Oxford University Press.

Gagliani, G. (2003). Terziarizzazione precoce e sentieri di sviluppo delle regioni italiane nel secondo dopoguerra: un'ottica internazionale. *Rivista economica del Mezzogiorno, 7*(1–2), 181–208.

Gambuzza, M. (1986). Gela: patologia o spontaneità urbana? In G. Amata, D. D'Agata, & M. Gambuzza (Eds.), *Inquinamento e territorio. Il caso Gela* (pp. 65–81). Cuecm: Catania.

Garcìa, M., & Karakatsanis, N. (2006). Social Policy, Democracy, and Citizenship in Southern Europe. In R. Gunther, P. N. Diamandouros, & D. A. Sotiropoulos

(Eds.), *Democracy and the State in New Southern Europe* (pp. 87–137). Oxford: Oxford University Press.

Ghimire, K. M. (2018). Path Dependence. Abnormal Times and Missed Opportunities. Case Studies of Catastrophic Natural Disasters from India and Nepal. *Law and Development Review, 11*(1), 31–76.

Ginatempo, N. (1976). *La città del Sud. Territorio e classi sociali.* Milano: Mazzotta.

Gordillo, G. R. (2014). *Rubble. The Afterlife of Destruction.* Durham, NC: Duke University Press.

Gotham, K. F., & Greenberg, M. (2014). *Crisis and Redevelopment in New York and New Orleans.* New York: Oxford University Press.

Gucciardo, G. (1996). Regolazione sociale e abusivismo edilizio: il caso della Valle dei Templi di Agrigento. *Meridiana, 26–27*, 247–276.

Hardin, G. (1968). The Tragedy of the Commons. *Science, 162*(3859), 1243–1248.

Harvey, D. (2010). *The Enigma of Capital and the Crises of Capitalism.* Oxford: Oxford University Press.

Heidegger, M. (2008). *Being and Time.* New York: HarperPerennial.

Holston, J. (1989). *The Modernist City. An Anthropological Critique of Brasilia.* Chicago: The University of Chicago Press.

Iovine, S. (2016). *Ecocriticism and Italy. Ecology, Resistance, and Liberation.* New York and London: Bloomsbury.

Jacobs, J. (1961). *Death and Life of the Great American City.* New York: Vintage.

Lapavitsas, C. (2013). The Financialization of Capitalism: 'Profiting without Producing'. *City, 17*(6), 792–805.

Lascoumes, P., & Nagels, C. (2014). *Sociologie des élites délinquantes. De la criminalité en col blanc à la corruption politique.* Armand Collin: Malakoff.

Lefebvre, H. (1991). *The Production of Space.* Cambridge, MA: Blackwell.

Leibfried, S. (1992). Towards a European Welfare State. In Z. Ferge & J. E. Kolberg (Eds.), *Social Policy in a Changing Europe* (pp. 245–279). Boulder: Westview Press.

Licordari, M. (2017). The Importance of Seismic Legislation and Reinforced Concrete in the Reconstruction of the Cities of the Strait of Messina. *Conservation Science in Cultural Heritage, 16*(1), 237–276.

Lo Curzio, M. (1985). Ricostruzione urbana e piano Borzì. In L. G. Di Leo & M. Lo Curzio (Eds.), *Messina, una città ricostruita. Materiali per lo studio di una realtà urbana* (pp. 19–28). Dedalo: Bari.

Longo, P. (1933). *Messina città rediviva.* Messina: La Sicilia.

Lupo, S. (1987). Tra società locale e commercio a lunga distanza: la vicenda degli agrumi siciliani. *Meridiana, 1*, 81–112.

Mingione, E. (1991). *Fragmented Societies: A Sociology of Economic Life beyond the Market Paradigm.* Oxford: Blackwell.

Mortara, G. (1913). Messina: Come Vive. *Giornale Degli Economisti E Rivista Di Statistica, 24*(3), 250–298.

Nixon, R. (2011). *Slow Violence and the Environmentalism of the Poor.* Cambridge: Harvard University Press.

Noto, A. G. (2008). *Messina 1908. I disastri e la percezione del terrore nell'evento terremoto.* Soveria Mannelli: Rubbettino.

Paci, M. (1984). Il Sistema di welfare italiano tra tradizione clientelare e prospettive di riforma. In U. Ascoli (Ed.), *Welfare State all'italiana* (pp. 297–323). Bari and Rome: Laterza.

Parrinello, G. (2015). *Fault Lines. Earthquakes and Urbanism in Modern Italy.* New York and Oxford: Berghahn.

Polanyi, K. (1944). *The Great Transformation.* New York: Farrar & Rinehart.

Redazione. (2013, July 22). Il risanamento a Messina: una telenovela lunga 20 anni. *IMGPRESS.* Retrieved from https://www.imgpress.it/politica/74299/.

Redazione. (2018, May 4). Messina, non è un paese per giovani. I numeri dell'emigrazione. *Letteraemme.* Retrieved from http://www.letteraemme.it/.

Regione Sicilia. (1990). Legge Regionale n.10/1990. *Interventi per il risanamento delle aree degradate di Messina.*

Regione Sicilia. (2002). Legge Regionale n.4/2002. *Interventi per l'accelerazione e il completamento del risanamento della città di Messina.*

Rixen, T., & Viola, L. A. (2014). Putting Path Dependence on Its Place: Toward a Taxonomy of Institutional Change. *Journal of Theoretical Politics, 26*(3), 1–23.

Saladino, S. (1966). Italy. In H. Rogger & E. Weber (Eds.), *The European Right. A Historical Profile* (pp. 208–260). Berkeley and Los Angeles: University of California Press.

Salvati, M. (1994). Da piccolo borghesia a ceti medi. Fascismo e ceti medi nelle interpretazioni dei contemporanei e degli storici. *Italia contemporanea, 194*, 65–84.

Saraceno, C. (1994). The Ambivalent Familism of the Italian Welfare State. *Social Politics, 1*, 60–82.

Sassoon, D. (1997). *Contemporary Italy. Economy, Society and Politics since 1945.* London: Longman.

Scott, J. C. (1998). *Seeing Like a State. How Certain Schemes to Improve the Human Condition Have Failed.* Yale: Yale University Press.

Sgroi, E. (1978). Lavoro e non lavoro: mercato del lavoro e dipendenza sociale. In M. Centorrino (Ed.), *Dipendenza e sottosviluppo nell'area dello Stretto. Un contributo all'analisi del dualismo meridionale.* Milano: Giuffré.

Sibley, D. (1995). *Geographies of Exclusion. Society and Difference in the West.* London and New York: Routledge.

Singelmann, J. (1978). *From Agriculture to Services. The Transformation of Industrial Employment*. Beverly Hills: Sage.

Sutherland, E. H. (1949). *White Collar Crime*. New York: Holt, Rinehart & Winston.

Tonelli, A. (2013). Fascismo e classi medie: un dibattito storico ancora aperto. *Società Mutamento Politica, 4*(7), 115–128.

Tyrrell, I. (2010). *Reforming the World. The Creation of America's Moral Empire*. Princeton: Princeton University Press.

Villani, L. (2012). *Le borgate del fascismo: Storia urbana, politica e sociale della periferia romana*. Milano: Ledizioni.

CHAPTER 5

Working and Dreaming at the Margins of the City

5.1 A Study on the Present-Day City

The previous chapter investigated the origins of the polarized social order that has emerged in the city of Messina over a long period of time. From the disastrous earthquake in 1908 to the reconstruction process led by Fascist regime, passing through the "slumization" of the city and the "Fordist" building policies of the second post-war era. And, finally, the present neoliberal policies.

Within the structural framework described in Chap. 4, characterized by a structural and permanent economic depression whose modern origin can be traced back at least to the 1950s, several thousand people still live in substandard housing conditions, in many ways similar to those found in the aftermath of the 1908 disaster—while their number alludes to decline only slowly.

Furthermore, the division of housing in Messina corresponds to a particularly evident stratification. For an understanding of this particular stratification, it is useful to read an excerpt by Ginatempo, written more than forty years ago, in which the author notes that:

> The economy controlled by these bosses [mafia, Christian Democratic Party, building, monopolies, local middle-class], based, above all, on public spending and on building, became a surrogate for industrial development. And so, we find only unproductive investments, purely to manage the economic situation as it is, in order to create stagnation and favor speculation.

D. Farinella, P. Saitta, *The Endless Reconstruction and Modern Disasters*,
https://doi.org/10.1007/978-3-030-19361-4_5

> In fact, these cities are, primarily, huge reservoirs of overpopulation "latent and stagnant" (Marx) [...] The labor market in southern cities is based precisely on this marginal working class that is utilized in temporary jobs, especially building and public works, or in microscopic businesses that are perpetually in crisis, working from home or informal work linked to the service sector. (Ginatempo 1976, p. 68)

It is a description that, on the one hand, tracks the source of a social division and, on the other hand, faithfully describes the profile of the actors active in the field. Today, the inhabitants of the shacks and public tenements are engaged in exactly those occasional activities that Ginatempo wrote about decades ago. As we have shown in the previous pages, the recent history of the city is marked by a substantial continuity in marginality, which involves the public sector and broadens the base of the individuals subjected to the precariousness of living conditions.

As we have shown, this is an order that hinges essentially on the division between a narrow bourgeoisie, a relatively small middle class, generally dependent on the State, committed to the public sector and private income, and a much wider class of individuals, variously occupying subordinate positions. Some of these latter are at the center of this part of the study.

This chapter focuses on these marginal workers and contains the findings of different research projects carried out between 2001 and mid-2017. During these years, the intense research activity conducted by the authors has made it possible to analyze the city from different standpoints (i.e., from the position of the precarious workers in the construction sector and the public sector as well as from the perspective of the slum-dwellers). The studies were undertaken by means of an extended observation period, within a space, moreover, that is "familial" for being home to both authors.

An initial exploratory ethnographic study was conducted at the end of 2001 and the beginning of 2002 (three months), in the dilapidated Gazzi neighborhood, where new social housing is mixed with the "temporary houses" built by Mussolini (and still called the "Mussolini houses"), and the "Garufi houses", a group of houses that surround "Villa Garufi", along via Gaetano Alessi. At the end of the nineteenth century, this area of cultivated land was owned by the wealthy Garufi family. Annexed to the main villa, known emphatically today as "the castle" by the locals, there were buildings that housed the servants and laborers, and animal stalls.

These groups of buildings have never been restored or modernized, and over time, they have become makeshift homes for the poorer citizens.

After an initial quantitative analysis of the structure of the economic fabric (2007), research was carried out on a population of precarious workers with low-level qualifications in the health sector—hospital porters (material was collected during the year 2008–2009); at the same time, the other author conducted research into a population of building workers (2011–2012). During the period between 2011 and 2012, some in-depth studies were carried out on a deprived suburb in the Southern part of the city, particularly in the area of S. Lucia-S. Filippo, also focusing attention on the adjacent areas where many residents live in social housing and shanties. The conduction of this segment of the study was favored by the fact that one of the two authors lives in this part of the city.[1]

The combination of residency in the area and the exploratory ethnographic studies carried out in Mangialupi-Gazzi (2001–2002) and Santa Lucia-San Filippo (2011–2012) made possible the direct and longitudinal observation of some of the processes under scrutiny.

These two ethnographic studies were also made possible thanks to the fact that one of the authors was involved in collecting data for the national Census, in both 2001 and 2011. Thanks to this fortuitous situation, this author was able to gain access to two different parts of the city where there are shacks, low-income and social housing, thus creating contact with the people being observed as the basis for this ethnographic study. There was also the opportunity to visit the shacks and social housing in other parts of the city, since many of the interviews carried out with the precarious hospital porters and their families took place in their homes, in many cases, in areas such as these.

Finally, in 2017, one of the authors joined a political collective—a "trade union"—which practiced the occupation of public buildings as a way of fulfilling the housing needs of evicted families.

All of these studies used qualitative methods: in particular, in-depth interviews (fifty for the research on precarious public workers and thirty-five for the research on building workers) and ethnography in the case of the collective. Further materials were obtained by means of participation in demonstrations, meetings, and through the collection of public documents (articles, photographs, and comments available on social media). In addition, several months of observation were carried out in the provincial headquarters of a trade union that represents building workers (Fillea-Cgil).

Further opportunities for observations in the poorer neighborhoods situated in the southern part of the city (Valle degli Angeli and Villaggio Aldisio in the Gazzi area, Villaggio UNRRA, Villaggio CEP, Santa Lucia in the *Torrente* San Filippo area; see Fig. 4.4 in Chap. 4) were made possible thanks to the particular social position of one of the two authors, who has lived in and frequented many of these areas since early childhood.

As discussed elsewhere (Farinella and Irrera 2014, pp. 198–199), far from ignoring the fact of the author's proximity to the research object, as a native of the area, and familiar with the distinctive traits, registers, situations, ways of doing things, and speaking, this was seen as something that needed to be solved in terms of the reflexivity of the research. There was, in fact, a real awareness that privileged access to the decodification of the place, its dialect, gestural expressiveness, and local stories should be reflexively problematized in an open dimension that could be negotiated between two extremes: on the one hand, the opacity of the ethnographic object (or the native informant) and, on the other, the transparency that belonging to the social actors' context might seem to guarantee (both in the empirical research and in the writing). In order to give an explanation of this intermediate dimension, where the observer's role is negotiated, it was important to show (retrospectively) that this was possible only by bringing common cultural resources into play—or allowing them to be recognized as such during the interaction—to show these precarious workers that some of their main "definitions of the situation" were being understood, and that there was a willingness to enter symbolically into their "game", in other words, to understand them politically by gaining their trust. Our awareness of this problem, during the interpretation of our ethnographic data, led us to reflect on the tension between "taking" information "for granted" (because culturally learned), which would condition our access to the empiric material and other tools that the ethnographic method offers. This meant leaving the question of the translation/mediation/elaboration "open", as far as both the empirical research and subsequent ethnographic writing were concerned. By introducing these elements of reflexivity, we were questioning and problematizing our *belonging to* the context. This did not involve introducing ethnographical distance, but reflecting in-depth on the epistemological and political terms of our *possible* closeness "through" the practice of ethnography.

This joint research aims to highlight the ways in which subjects belonging to subaltern classes have made use of the gaps left open by local powers, to access forms of social mobility, even if they are fragmented and unstable. Moreover, it is an exploration of their "moral worlds", in consideration of the fact that, as we have shown over the course of the previous chapters, they are both a highly disadvantaged category and a severely stigmatized "group". They are, in other words, a category of people who not only face material problems stemming from an extremely classist organization of society but which is also blamed for the materiality of the experienced condition, being depicted as a "group" located at the margins of civility and endowed with its own specific culture. Regardless of the fact that this "pseudo-group", linked only by class membership, is not made up of people who belong to one specific, separate "culture", religion, or ethnicity, the people said to belong to this category are, nevertheless, imagined, perceived, and told by those who hold the "power to nominate", as an almost indistinct entity—almost a "race" apart, the *zalli*. Thus, while we think that subalternity and class differences exist and must be interpreted, we are also aware of the risks of contributing to the "orientalization" of a segment of society. For all these reasons, it is important to highlight that this is not a study on the "culture" of a homogenous group, but an investigation into class inequality and the ways existing asymmetries of power create their "others" and their "counterparts", together with those opportunity structures that delimit the course of action for the members of the lower classes. Subsequently, this is also a study on the worldviews that stem from, and are shaped by, the fact of occupying a subaltern position on the social ladder.

Hence, beyond being an inquiry into the historical continuity of class formation in a city, this is also a study of those "working poor" who, since the beginning of the new millennium, have found themselves under the lens of many international studies, and that, in the Italian and southern context, have not only had to deal with the restructuring of the labor market and the welfare system common to many countries but also with the peculiarities of national and local systems. These include the delay or suspension of public infrastructure works, the contraction of the housing market, the massive use of subcontracting, and the consequent tendency of employers to pay little or nothing, omitting even the payment of contributions, and also the frequent use of contractual formulas that have no bearing on the reality of employment relations.

Through biographical analyses, this study provides some reflections on the subaltern condition in the city of the Straits and, to some extent, in southern Italy as a whole. The studied biographies appear individualistic and liminal, suspended as they are between the legal and illegal, and seen alongside an order that can ultimately be defined as *mafioso*. However, the study puts forward the idea that these existential paths and the connected visions should be seen as intimately political processes—aimed at moderating asymmetries and structural injustices—as well as the outcome of emotional processes characterized by the systematic frustration of aspirations and desires.

As we have tried to suggest over the course of the book, the perspective presented here aims also to emphasize the fact that, although parts of the South are seemingly going in different directions with respect to other parts of the country or of Europe, this is, more than the result of internal features, the outcome of complex forms of convergence between "core" and "peripheral" elites that date back to the Unitarian process—a convergence of interests aimed at specializing the interior territories and assigning precise functions within the domestic division of labor.

Moreover, the eternal precariousness of Messina and the South is understood here, in particular, as a signal for the end of the welfare state and Fordism, and the advent of a late liberalism (or, if you prefer, neoliberalism), globally characterized by the same inequalities and forms of instability that have historically identified this part of Italy.

5.2 Getting by in an Informal World

The research segment, which took place between 2008 and 2009, was carried out with a group of health-care assistants from the Polyclinic hospital in Messina, who had been working precariously and struggling to have their right to job stabilization recognized, following the implementation of the Prodi government's financial policies. The interviewees were all fifty years old and above, and they had been listed as unemployed since the 1970s, but they had worked reasonably regularly in the precarious public sector since the 1990s, following their qualification as "health-care assistants" (previously called hospital "porters"). The main data was collected by way of fifty in-depth interviews about the life stories of unskilled precarious temporary workers and their families and other ethnographic materials (see Sect. 5.1). More specifically, there were forty-two in-depth interviews with health-care assistants (thirty-one women and eleven men); five interviews with other family members (of which three husbands, one wife, and one sister, some of these interviews are double: in the same recording there are workers and their relatives); and, finally, six in-depth interviews with other key actors (four trade unionists, one politician, and the general director of the State hospital).

From these data, we concentrated on the organization of the workers' economic lives, and that of their families. The workers were "marginal", not necessarily because they were poor, but because they were trapped in a condition of existential precariousness which seemed to get worse and worse (Bourdieu 1999).[2] The context was one of families of origin, with a head of the family who is underemployed, with large families, precarious living conditions (often marked by the fact that they lived in a shack or in post-earthquake accommodation as they waited, seemingly forever, for more permanent housing), a lack of school qualification (many of them only had the compulsory school attendance certificates gained at evening schools) and work skills, all of which eliminated any possibility of social mobility and forced them to adapt to what the fragile local system had to offer: bad work, labor-intensive and low-production manual activities, affected by Baumol's cost disease,[3] which did not allow for the accumulation of salaries that could sufficiently support a family. These people were surrounded by the "informal", within which they had developed survival tactics, which acted as stabilizing factors to help contain their social insecurity. They invented social practices, some successful, others less so, within their territory, often making use of relational networks, and by imitating, by inventing, by connecting themselves to the particular-clientele channel, by opportunistically exploiting any of the benefits available from both institutional and informal structures. This being the legacy and the outlet, as we have seen at the beginning of this chapter, of a governmental "technology" developed toward the end of the 1910s, perfected by Fascism, and enhanced in the post-war era.

We also focused on the "survival" tactics they put in place in order to resist and act against social marginalization and to create opportunities for social mobility. According to De Certeau (1984, p. XiX), tactics are different from strategies:

> I call a 'strategy' the calculus of force-relationships which becomes possible when a subject of will and power (a proprietor, an enterprise, a city, a scientific institution) can be isolated from an 'environment'. A strategy assumes a place that can be circumscribed as proper (propre) and thus serve as the basis for generating relations with an exterior distinct from it (competitors, adversaries, 'clienteles', 'targets', or 'objects' of research). Political, economic, and scientific rationality has been constructed on this strategic

> model. I call a 'tactic', on the other hand, a calculus which cannot count on a 'proper' (a spatial or institutional localization), nor thus on a borderline distinguishing the other as a visible totality. The place of a tactic belongs to the other. A tactic insinuates itself into the other's place, fragmentarily, without taking it over in its entirety, without being able to keep it at a distance. It has at its disposal no base where it can capitalize on its advantages, prepare its expansions, and secure independence with respect to circumstances. The 'proper' is a victory of space over time. On the contrary, because it does not have a place, a tactic depends on time—it is always on the watch for opportunities that must be seized 'on the wing'. Whatever it wins, it does not keep. It must constantly manipulate events in order to turn them into 'opportunities'.

It is important to make a clarification here. The choice to term these practices as tactics is not linked to the fact that these people are governed by contingent, spontaneous, and chaotic actions that are unplanned and directionless. On the contrary, as seen in this chapter and in the next, many of the practices adopted by these subjects are guided by rationality, are designed to reach a specific goal, and pursued by calculating the benefits and opportunities, showing how the subjects' individualistic and economic actions are capitalistic in nature. However, these practices are fragmentary and unpredictable because they are intrinsic to the place where they originated. They represent daily attempts to make use of the ordinary, or better, to invent "other" actions or possible opportunities, or find temporary systems, while being perfectly aware that these are only partial solutions that are often destined to fail. Metaphorically speaking, we might say that these tactics make it easier to find temporary solutions, or "patches", to help mend the fabric, but fail in their main objective to make the dress fit better, or to make it look new by adding more material.

Furthermore, these tactics are always ambiguous, ambivalent, and in constant tension: on the one hand, they are crushed by the need to guarantee security to the family, to protect them from social risk and from the economic insecurities of precariousness, and in this way substitute the worst of the shortcomings of the "modern" local welfare system by playing the role of stabilizing anchors. On the other hand, their implicit ambition is to search for ways to achieve existential stability that might also include forms of social mobility: finding a secure job, *for life*, or a dignified and stable home, are assurances for the future, for old age, that go beyond a simple survival strategy, and become a project for the future. This tension remains unresolved, but perfectly aware of the situation, the subjects and their families put into practice a form of existential do-it-yourself (DIY), whereby they use numerous chaotic tactics, many of which are redundant, copied, sketchy, certainly impracticable, and unrealistic. However, they are always worth trying because they might open up a universe of possibilities, of uncertain and ambiguous plausibility, which might prove to be an unexpected way to reverse marginality and, to different degrees, to increase their social, economic, and relational stability, for themselves and their families (Farinella 2013).

From their statements, it is possible to discern a universe of plausibility (rather than truth), where the subjects, their tactics, and the environment are at times plausible, truthful, or transformed, but in the way they are "expressed", they communicate a world of local relationships (Geertz 1983), successfully staged, in their use of words and metaphors, in the plots that unfold, in the characterization of the "cast" (the politician, the unemployment office, the usurpers, the victims, etc.) that seem to be straight out of the *commedia dell'arte* (Farinella and Irrera 2014).

Figure 5.1 shows the earnings that a family of marginal workers derive from the integration of different survival tactics. It is evident that these practices are dissimilar; many of them are simply ways of guaranteeing the survival and stability of the family in a context of high precariousness and

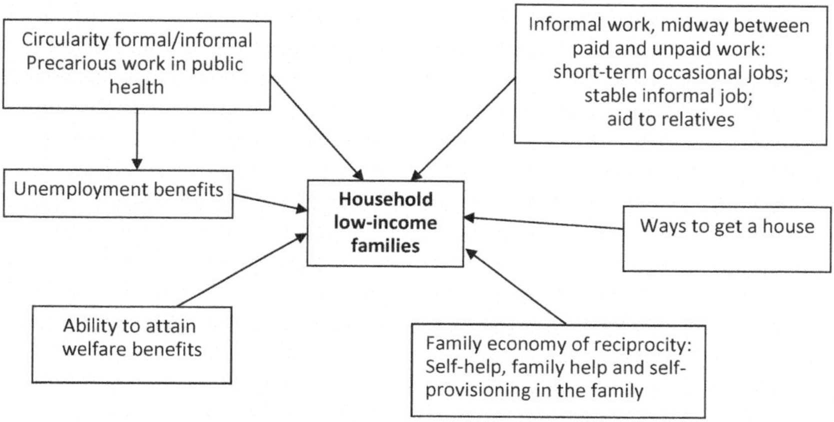

Fig. 5.1 Survival tactics: different stabilizing anchors to help contain social insecurity and marginality. Source: Our elaboration on the in-depth interviews

limited resources, others outline ways designed to gain social stability in the long term in order to improve social conditions (in particular, techniques for acquiring a "secure" job or a house). Depending on the situation, these tactics are used in a complementary and integrated way, or as an alternative or to compensate: they can therefore be simultaneously used in the same family or by one or more members of the family, in a complementary way, to supplement the total family income; otherwise, they can be used at different times and are transitory in nature, compensating or substituting often in view of temporary legislative opportunities that arise, but only for the brief period they are in force. Perhaps the most interesting aspects are the adaptability and flexibility of these tactics within the family context, and the fact that they can create opportunities to help the whole family unit (guaranteeing its survival and, wherever possible, a certain degree of social mobility). These are tactics that develop within the informal and cross over it, underlining exactly how the mechanisms of the informal economy function in lower-class contexts, as substantially highlighted by international literature since the 1970s (among others, Bagnasco 1986; Portes et al. 1989; Benton 1990; Williams and Windebank 2003; Leonard 1998, 2000; Portes and Haller 2005; Guha-Khasnobis et al. 2006; Coletto 2010; Saitta et al. 2013; Agarwala 2013; Boels 2016).

One of the main stabilizing factors was their ability to find work in a guaranteed sector. By "guaranteed" we mean a sector in which the worker is protected against job insecurity for their entire life, through the guarantee of a stable job, the certainty of obtaining an old-age pension, and health insurance in the case of work injuries or illness, and unemployment benefit in the case of job loss. In spite of the changes, in Italy, the sector that offers all these benefits and, in the case of a permanent contract, gives protection against dismissal, thus giving the guarantee of a job for life, is historically the public sector. This has resulted in the tendency, especially in the South, to seek a job in the public administration, which has created a progressive process of "meridionalization" in the public administration (meaning that there is a high prevalence of subjects from the south working as public employees. See: Cerase 1990).

Several studies have pointed out how choosing to work in the public sector rather than the private is based on individualistic motives (job security), rather than on interest in collectivity or in the public interest (Cerase and Farinella 2009). This individualistic motivation can be seen as the reason behind the clientelistic system of votes in exchange for jobs, as was seen in Chaps. 3 and 4, and how it plays the role of stabilizer of the social fabric in the South's public administration (Paci 1989; Trigilia 1992; Ferrera 1996), a role that has generated a sort of particularistic inclusion for a large part of the population with a medium-low income.

As emerged from the in-depth interviews, for the unskilled workers, the ability to get a job in the public sector, even a precarious one, was a primary objective. In the meantime, working in the building sector as a laborer guaranteed some form of security and some benefits, even if temporary, because, generally, laborers could obtain a temporary contract, while working as a worker in the private services sector meant working informally, without any security.

Given the difficulty of obtaining a full-time contract (especially after the reforms to the public sector during the 1990s, and the subsequent freeze in recruitment), precarious temporary work in the public sector (and for workers in the private building sector), with the possibility of alternating periods of stable work with periods of "black" work (despite the possibility of obtaining benefits linked to unemployment), became a sort of safety net against instability. Other forms of activity that were completely original and which seemed to create circularity between the formal and informal economy derived from this. This circularity had already been observed by Lomnitz (1988, p. 54) and Portes and Haller (2005, p. 409) who had noticed that it is actually the State's attempt to regulate that ends up supporting informality (as an unexpected by-product of state control).

The hospital porters, for example, work for several months of the year in the public sector (with job guarantees and formal contracts), but during the rest of the year, they work in the private sector market (as irregular workers, carrying out unspecialized work in the service industry or commerce, without any guarantees and only an informal agreement with their employer). From the interviews, it emerged that these workers consciously choose to remain irregular and to work informally in the private sector. They do so in order not to lose their right to unemployment benefit and, more particularly, the right to be "called up" for public sector work (a regular job in the private sector would annul the points accumulated at the unemployment office).

> At the age of eighteen. […] I started working as a babysitter [in the 'black' economy]. […] In this job as babysitter, I never had any contributions paid […] Quite rightly, she wanted to regularize my position, to be fair. It was my choice to work illegally, […] because I knew full well that if she regularized my position I'd lose everything at the unemployment office. […]

> Instead, in this way, I also had the chance of working at the Polyclinic hospital for three months (through the unemployment office). Of course, when I worked at the Polyclinic, it wasn't the same salary as I was getting from them! (Worker, Polyclinic hospital, woman, aged 48, married, 1 child, interview n. 11)

For the workers in the construction industry, there is also a circularity between the informal and formal. They often work for six months with regular, formal contracts and go on to claim unemployment for the remaining six months and continue to work illegally for the same company, while unfairly continuing to claim unemployment benefits. Within the building sector, there is also this circularity between the informal and formal. It is a mechanism that off-loads labor costs and company costs opportunistically onto the State and is more cost-effective both for the business owner (who can therefore avoid paying part of their labor taxes, which are very high in Italy) and for the worker, who compensates for any risks by "moonlighting" for a higher salary (made up of the salary from their informal work added to their unemployment benefit). In the following excerpt, a husband explains the complex mechanism of hiring and agreed layoffs which allow them to maximize their unemployment benefits:

> Interviewer—How are you going to manage in the future? Without contributions, without anything.
> G.T.—With a regular contract, I've got [only] eight years of contributions. […] I've almost always worked irregularly so as to get unemployment benefit. For example, I worked for six months … in the building industry, with this big company. Six months, with the minimum contributions to get the construction worker's security fund and unemployment benefits. Then you quit and you get three months of special unemployment benefit for the six months you worked, and automatically, you work and you get benefits too.
> I.—And they let you do that? They all gave you a salary?
> G.T.—Everything. That was the arrangement, because I used to get fewer stamps … when I had a regular contract, but then I got unemployment benefit. Then when the three months, six months were up, I had to do another year to get the six months. Three months' work and three months' unemployment. It was all a…
> I.—What, what? I don't understand. What do you mean a year? You worked six months…
> G.T.—So … to get unemployment benefit you need fifty-two weeks of stamps.
> I.—Ordinary unemployment or special?

> G.T.—Ordinary, ordinary for six months. Before it was six months, now it's eight months or ten months depending on your age. Before it was six months for everyone. You had to work for fifty-two weeks to get six months of unemployment benefit.
> I.—And so, 52 weeks were how much?
> G.T.—A year. It was a year.
> I.—Ah, right, right. A year.
> G.T.—You had to work for a year to get six months unemployment. At the same time as you applied and got six months' unemployment, Inps (National Social Welfare Institution) sent you three months of special unemployment which corresponded to 80% of your daily wages and three months of ordinary. Also, they gave you first the special and then the ordinary. I mean, first you got seven million Lire and then you got three. When you finished the seven million, you finished the first three months of the special and there were three months remaining, you could speak to the engineer, and get a regular contract for six months. Six plus three equals nine, because you had three over, you could go onto unemployment again, there was no need to wait a year, but just six months. But what did Inps do at this point? They no longer gave you the seven million and three were left aside. They gave you first the three and then the seven. So you had to get all six months. It's a question of mathematics. Get it? It was all a game of numbers. So when the unemployment expired, you had to start again for a year.
> I.—So, first you got three, and then you got seven. Then didn't you get the other three again?
> G.T.—No. You work a year to get six months, for the right to six months. Inps gives you three months…
> I.—And then, after you had these three special months, you got hired again for six months, and therefore you didn't get the three months of ordinary.
> G.T.—No, I didn't get that, I didn't get it. I left it there put to one side. After six months, I'd apply again and I had the right to get three months…
> I.—… of special. (Male, unemployed, estranged husband of a precarious worker at the Polyclinic hospital, aged 49, 1 child, interview n. 15).

The rotation mechanism between formal and informal work permits the workers both in public jobs and in the construction industry to increase the proportion of workers in guaranteed jobs, even if these are only temporary, thus generating an effect of job distribution.

Work in the informal sector creates a further stabilizing anchor which allows families to supplement their total family income (Mingione 1983, 1985), even though the stifling local work market and the large mass of the unemployed trap workers in bad work situations (a way of "safeguarding" themselves from excessive "exploitation" is to gain access to informal work through the help of friends and relations).[4] The informal division of the labor market is also characterized by a gender division: men's jobs include construction work (manual), commerce, and the service industries (unloading, assistant cook, bread-making and pastry assistant, delivery boy); women's jobs include services, the care sector, and cleaning (seamstress, babysitting, care of the elderly, domestic help). Irregular jobs generally fall into three categories: occasional odd jobs or one-day jobs, which are found, thanks to the help of friends; more stable but undeclared jobs, carried out concurrently with precarious temporary public sector work (in effect double jobs, both as employee and as owner); and lastly, collaborators in family activities such as bars or shops.

The contours of these activities are hazy and can vary between traditional *paid informal work*, to more complex mechanisms such as *unpaid community work*, based on reciprocal exchanges and self-help (among others Leonard 2000; Williams and Windebank 2003) that imply obligations of a symbolic nature rather than a monetary one (Bourdieu 1998, p. 93 passim). These might include short-term occasional jobs, giving relatives or friends a hand with their activities:

> I make do. The occasional odd job, now and again. [...] Totally irregular. But just odd jobs. [...] For a while I was working for my brother. I answered the phone at his shop. [...] Then he closed down [...] When the school year started, for two months I went to get books from the warehouse in Catania with a friend of mine who owns a stationer's. We'd get up at three in the morning, and got back at four in the afternoon. [...] Then I worked unloading lorries full of plants. There was this friend of mine, unfortunately he died four years ago, who worked with plants, and they brought these lorries full of plants which needed unloading [...] Whatever was available. By the day. (Male, polyclinic hospital, aged 50, single, interview n. 17)

> Either I didn't work, or I'd find some little job, like doing housework, or looking after some old woman. You know, just little things [...] I found a woman who I still go to and I just went to her house and did some housework. (Woman, polyclinic hospital, aged 54, married, 2 children, interview n. 6)

> No, I did a bit of private care with people who needed looking after, not always, for example a few nights with an elderly person in hospital, thanks to my colleagues who knew me and called me when there was care work to do, [...] I had quite a good round going, I was well-known and they would often call me. (Woman, polyclinic hospital, aged 41, married, 3 children, interview n. 10)

(I began working as a seamstress) when I was seventeen, [...] at first I was working for a dressmaker [...], then I set up on my own at home and did some sewing for my clients because I had some regulars. [...] I even managed to earn quite a lot. (Woman, polyclinic hospital, aged 44, single, interview n. 12)

My dad had a small bar and I worked there. [...] I just gave my dad a helping hand. [...] My dad supported me, I gave him a hand, he would give me something. Then in '94 they called me and I came to work here (at the Polyclinic hospital). [...] But when I wasn't working, before I came to the Polyclinic, I went to work for my dad and helped out. But we didn't do much business because it was a local area. (Male, polyclinic hospital, aged 44, married, 3 children, interview n. 16)

These forms of irregular work intertwine: one can simultaneously work in the public sector, run or collaborate in a commercial activity, and do the occasional odd job:

A.—[...] I opened up in the hardware business, a key center, where I cut keys [...] and I also sold fishing tackle [...] I've done lots of different types of work, odd jobs. [...] I also do photography, I'm still a photographer. I had this hobby of photography [...] and then I worked with this photographer, and I went to help them out and that's how I learned the job. I tried to pick things up [...] They would call me for jobs, weddings, christenings. I was their errand boy, you know! The bag carrier.
Interviewer—Was it always in black labor?
A.—Of course! Oh, always that, always. [...] Only the key business was regularized, it was the little odd jobs that were 'in black'. Then, I left the business because it was a big commitment [...]. But, in the meantime [...] I also worked some months here, at the Polyclinic.
I.—And how did you manage that, if the shop was registered in your name?
A.—It wasn't in my name. The shop was registered in my mother's name. [...] Yes, yes because I thought who knows who might get the job here. That's all. You know, if I was regularized I wouldn't be able to make (an application at the unemployment office). [...] So then I closed it down. [...] and decided just to do the four months at the Polyclinic and my photography. (Male, polyclinic hospital, aged 45, married, 1 child, interview n. 9)

Other stabilizing anchorages to contrast social insecurity come from the ability to attain welfare benefits and other support through the formal system, for example, linked to the status of "family" living in reduced circumstances, invalidity allowances for parents or children, one-off payments such as the "baby" bonus or free medicines and medical examinations, rent contributions, and so on. This use of public benefits to help informal and marginal workers cover costs and needs can be found in many other contexts. (e.g., see Agarwala 2013).

Finally, another tactic makes use of the complex reciprocal economies that informally arise in the family unit, in particular, in forms of intergenerational redistribution, often when the family of origin plays an important support role, such as helping to look after grandchildren, the purchase of consumer and durable goods, and sharing resources.

5.3 The Public Sector and the Circularity of Formal and Informal

As already mentioned, in a context where there is little in the way of a "future", public sector work, with the stability it brings, is an opportunity for some sort of "arrangement": on the dimension of secure/insecure (Castel 2017), (public) employment contrasts with (private) odd jobs, which belong to the sphere of making do, and which wear people down in the long run:

> I was sewing full time just to support the family [...] And for me, this sort of precarious work [...] has consumed my life [...] I love sewing [...] more than anything else [...] But it doesn't give you what this sort of work (as an health-care assistant) can give you, because here you have a pay packet, you get holidays, it's a more secure situation. [...] I can look forward to living more comfortably in my old age. (Woman, polyclinic hospital, aged 56, divorced, 1 child, interview n. 1)

> Before this job, I worked as a babysitter [...] for a doctor, who knew me so I worked in black. [...] I've always tried to keep busy. I did some ironing, three hours at this woman's house. But you know, it isn't that it's not rewarding or that it isn't an honest job, but it's different working at the Polyclinic, having a thirteenth salary, having paid holidays, if you're ill, [...] you have paid sick leave. Lots and lots of things that I had no right to when I did occasional work because I needed the money. (Woman, polyclinic hospital, aged 44, divorced, 2 children, interview n. 2)

If, therefore, secure work "passes by" from state sources (Cassese 1993), a credible strategy is to try and "jump on the passing train" (as said by an interviewee). But it is not as simple as it sounds. The workers in this study proved to be doubly marginalized: in addition to being "peripheral" to the local economic system (in which they represented the work "surplus"), they were equally so in the particular clientelistic system. For over twenty years since their registration at the unemployment office, they had been unable to gain access to stable work, even though this was the same time period during which public sector jobs in the South increased considerably. Even though they are an integral part of the system, they are, in fact, the least important components, trapped in a position of weakness and subordination that comes from being excluded and from wanting to be a part. In their continual search for even the smallest contact between their informal world and the system of "formal" guarantees (whether it be a friend, a politician, a person of importance, legislation, an office, a relation, a qualification, a category, an illness, a certificate, etc.), they described relationships, social space, and representations in which they were "exploited" by the system. But in the "portrayal" of themselves as victims, an active subjectivity emerges in the tenacity that helped them to endure and accumulate scenarios of missed opportunities; in the complexity of the strategies they invented in order to force entry into the public space and opposed rejection; and in the original "explanations" of facts and laws that they "gave", which often "lent drama" to a simplified and misunderstood universe, but which defined a situation of limited certainties to which they could cling:

> The older ones who had over eighteen years (of unemployment) were supposed to be, at least in theory, placed in a job for at least three months a year. There you are. And that's what Law fifty-six stipulated![5] (Woman, polyclinic worker, aged 63, married, 2 children, interview n. 3)

In the stories of missed opportunities, whether due to inexperience or naivety or to the "misfortune" of not finding the right support, or because a promise was not kept, the subject is never passive but becomes active and organizes himself or herself in order to obtain the necessary prerequisites (an educational qualification, a certificate, a "credit") on which to base their rights to what they are "due":

> The first job they called me for, well, my father wouldn't let me do it. [...] Garbage collecting. My father said no. [...] My mother worked for years for the Finance police, at that time she was a seamstress. [...] Then, my father, out of jealousy, made her leave. [...] My mother might have had this chance to get her children into employment. [...] there, there was the Colonel's

wife who used to say ‘Maria, just make sure your children get through school and then, you’ll see, you can come here, I’ll sort it out, I know what I have to do’. That was how you got in then [...]. (Woman, polyclinic hospital, aged 48, married, 1 child, interview n. 4)

It’s not that I haven’t looked for some sort of “way”, it’s just that all the doors were closed. I’ve never found the right key. [...] I once did an entrance exam and I did well. [...] when the classification list came out [...] I was among the first. [...] the clerk said “Aren’t you happy!? If you’re on the list then you’ll be called soon”. [...] Then I met a colleague of mine, who was already working and he had been below me (on the ranking list) [...] So I went back (to the unemployment office) and this ranking list had disappeared, it had just vanished. [...] I’d been on that list. Then someone took my place, and because they could see that I was a decent person, they said “just keep quiet, otherwise it’ll end badly”. [...] I’d also got a lawyer to take up my case. I even wasted money on the lawyer, because everything had been documented and I had the proof, but then I got this phone call at home and I had to drop everything. (Male, polyclinic worker, aged 55, married, 2 children, interview n. 13)

Our son has had to emigrate, but we tried thousands of different things so that he wouldn’t have to go [...] for example, I went to a politician, I helped him with his election campaign, not just me, the whole family, and it went on forever. When he was 18, my son got his radio operator’s license in Genoa. He left when he was just 18 and two months old, and he was the youngest of all [...] by the time he was 20, he was a sergeant with a license and with navigation experience. [...] So nothing, my son had to go away because they didn’t keep any of the promises they had made [...] that they’d get him a job here [...] on the ferries (in the Straits of Messina) which cross the strait, [...]. The most important thing was to have two years’ navigation experience, which my son had, and that he’d done his military service in the Navy and not on land, because my son had been on land, but we got him changed to the Navy. So, we made enormous sacrifices and all for nothing. (Woman, polyclinic hospital, aged 50, married, 2 children, interview n. 1)

Wife—In 1976, I don’t remember when, they announced a competitive exam at the local council, for the job of garbage collector. [...] And so my husband applied for the exam. There was this person, and he had built him a house for free, he’d built him this villa for free.
Husband—Me and my brother. On Saturdays and Sundays.
Wife—Saturdays and Sundays. They worked for free! They were supposed to give us the house and they didn’t. So then I said, ‘Mr X, at least get us a job!’. He said ‘when the exam for garbage collector comes up, don’t worry,

> I'll get you in!' […] He's sorted out millions of people! So, when the exam came up, and he went to see him and said 'Mr X the exam has come up'. He says 'do it but you have to pass the written test, then I can do something for you at the oral test, but if you don't pass the written paper I can't do anything'. So, at the written exam he sat next to this boy who had a high school diploma…
> Husband—Because, you know, I haven't got any school qualifications!
> Wife—He never went to school! He got the fifth year elementary qualification at evening classes! And so he passed the exam because he copied from that boy! […]. 'Look Mr X, I passed the written exam'. 'Oh, okay, then there won't be any problems with the oral. When they call you for the oral exam, just let me know'. They were supposed to call him for the oral, but in May he (the politician) had a heart attack and died! (Woman, polyclinic worker, aged 58, married, 2 children, interview n. 19)

The longed-for possibility for a link to the institutional system only came about during the 1990s, when the effects of the first laws for the flexibilization of the job market (Dlgs 29/93, Article 36) made it easier to find precarious work for people with low-level qualifications—where the job proposal comes directly from the unemployment office. In this way, it is possible to obtain an initial three-month contract as a general health worker, at the end of which they gain the qualification of "health-care assistant". And it is this latter that becomes part of a "category of established representation" (Pieretti 1991, p. 34), thanks to which it is possible to gain access to public benefits: they are put on a short list of workers that the health sector regularly uses due to chronic staff shortages. And it is through this interweaving of regulations that the formal begins to produce the informal. Regulation 56/1987 states that a precarious worker maintains their "unemployed" status at the unemployment office and their position on the ranking list, if the job they are doing does not exceed a period of four months in the calendar year. This is the start of a mechanism of rotation between the health-care assistants in which each of them works for three months of the year in the public sector, and spends the rest of the year trying to "get by" in the informal sector, so as not to lose their right to be called. In this way, precarious public sector work, unemployment benefits, and informal work become indelibly intertwined.

> I worked for over twenty-five years in the fruit and vegetable market. That is, it was always temporary work, always there, and in the meantime, in 1992, we began […] getting the first jobs as social-health operators or, at

least, as general manual workers. In 1994 they recognized this qualification and from there we started working as social-health operators. [...] As far as work was concerned, I never had any certainties. [...] from 1994 onwards we did three months and, because we knew then that, in theory, you ended up with a place here, it was just a question of time, so as not to lose this, you did both one and the other. I got up at three, and went off duty at eleven. I came here, I did my afternoon shift and went off duty at eight. And all this to stay on a certain course, on a certain list, so as not to refuse the three months' work. (Male, polyclinic hospital, aged 49, married, 3 children, interview n. 14)

No wonder, then, that a sort of "club effect" is created, linked to the public good of the "job opportunity". When workers see how advantageous it is to "belong" to the group of "health-care assistants", they all try to find a way to "get in". In this way an access "market" is created, leading to an increase in the number of beneficiaries, and made possible by the fact that the prerequisites for access are never "objective", and change depending on what is needed: "it's like a bus, if it's full, either you get off and I take your place, or we squeeze up so I can get on too" (male, polyclinic hospital, interview n. 15). This is also why there has been a boom in regional and private courses, which result in the much sought-after qualification, and also an increase in the number of qualifications issued by private and quasi-public care institutions, where it is possible to get around employment restrictions by acquiring the prerequisites, either through acquaintances or recommendations or through being willing to "pay" to be regularized or through being in the right place at the right time. Lastly, there is also a possibility of being "called" as a manual worker by the employment office itself, on the basis of the "real" or "supposed" need for additional personnel, who, after three months, obtain the qualification and also a place on the ranking list.

The employment office has ruined us, in Messina, with this qualification, nobody got in. [... But] there are other hospitals: Patti, Mistretta, Taormina [...] So, what did they do? In the absence of porters, they called the manual workers [...] once they had got in, when the time came to leave, they no longer came out as manual workers, but as porters, health-care assistants. (Woman, polyclinic worker, aged 54, married, 3 children, interview n. 6)

There were external courses and private fee-paying courses. [...] the regional ones only took a few people [...] So, everybody did the fee-paying courses! Two thousand euro each, they must have made a lot of money! (Male, polyclinic hospital, aged 44, married, 3 children, interview n. 16)

> Some people, the crafty ones, went to the clinics [...], how much does it cost the State to regularize someone for three months? Three million Lire? They paid three million, they were given a certificate saying they'd worked for three months as an health-care assistant, as a porter, and so they got back in! (Male, unemployed, estranged husband of a precarious worker at the Polyclinic hospital, aged 49, 1 child, interview n. 15)

In the various accounts of the ways in which the *outsiders* go about acquiring a qualification, there is always a double register: on the one hand, the *insiders* are the "victims" of the others' abilities to bend the rules in the system of formal opportunities, through irregularity and informality; on the other hand, there is an unspoken awareness that these "others" are also victims themselves, and that on more than one occasion, they have found themselves in a borderline situation, where it is very difficult to distinguish between good and bad, true and false, legal and illegal. In fact, each respondent had tried hard to enter the category in one way or another, either to stay or to get ahead (by obtaining school certificates at evening classes, by downgrading their own qualifications at the employment office in order to be called up, by separating from a spouse, by increasing the size of their family):

> I've always worked. I haven't done any school courses. My wife went to evening classes instead of me, so [...] well, I started [...] doing the first three months here. (Male, polyclinic hospital, aged 49, married, 3 children, interview n. 14)

> There were some people who declared anyone as a dependent, relations, half-relations. All sorts of strange relatives appeared, and they always got a few months. (Male, polyclinic hospital, aged 50, single, interview n. 17)

> There was always this favoritism for people who had been recommended [...] Because I have an invalid son, [...] I had to get a legal separation from my wife so that she appeared as a single mother with one son, who was entitled to double points, 24 points, to get more points than people with two or three children, who get 6 points each. [...] with four children they throw in an extra 24 points, and so my wife has finished working! That's why we did certain things. I started to understand the ins and outs, and when I used to go there, I had to behave like a real delinquent. I would go in, kicking the door, threatening, shameless [...] But it wasn't only me! Lots of people. To these threats they'd reply [...] well [...] "what position is your wife on the list?" "Thirtieth". "Well, she just gets in then!" [...] Who are they kidding! [...] after all my angry explosions, they put her back on the list again. (Male, unemployed, estranged husband of a precarious worker at the Polyclinic hospital, aged 49, 1 child, interview n. 15)

They were individuals who used their intelligence from a utilitarian perspective, with a cost-benefit analysis, which also took into consideration the risks involved in "borderline" actions. Even so, it was actually the employment office, which, in the various accounts, is represented in public space as an "antonomasia", which made this borderline uncertain and negotiable: laws and ranking lists became beyond words, ephemeral, incomprehensible, and changing from one day to the next.

"Getting bread with your card", "the place of hope", "the madhouse"—were just some of the metaphors used to describe the entropy of the employment office as public space, where "they never give you a proper explanation" (Male, polyclinic hospital, interview n. 18), "whoever has power commands" (Woman, polyclinic hospital, interview n. 7), and "whoever wields the power knows what to do. You find out about things because someone tells you 'that's how it works', but no one will ever tell you the truth in an honest way" (Male, polyclinic hospital, interview n. 19). In comparison with those in "power" who deal with specific knowledge and control information, the precarious worker is always one step behind:

> W1—In my documents, I appear as the head of the family [...] because I was once told "if you make yourself the head of the family you'll get more points". [...] So I make myself head of the family. Then, I go to the town council, and I just go crazy [...] I go to the employment office [...] I take all my documents, "no, madam, we don't need this". "What do you mean, you don't need it?" "No, because as far as we're concerned, your husband could even be Saddam Hussein. This is an individual application". "What? You told me you wanted to know my husband's income, up until yesterday!" "Yeah, but now we don't". Then you go again and this time they tell you they want to know the income. You see? So, they just do your head in. [...] They just put a law across any way that suits them. [...] Today it's like this, tomorrow it isn't. [...] It's like this all the time! And they say "hey, madam, what can we do about it! Try again, maybe next time you'll have more luck". [...] "But how come Mrs so and so who has only done the months four times is getting in, but I'm not getting in, and I've done them thousands of times?" you know, for example. "Ah, but that woman has her children, her husband, her grandmother, grandfather, grandchildren, who are all dependents" [...]
> W2—Nowadays my eyes are wide open. Once upon a time they were closed!
> W1—Yeah, but we opened them too late! (Dialogue between women, polyclinic hospital) (W1—Woman, polyclinic hospital, aged 56, married, 2 children, W2—Woman, polyclinic hospital, aged 52, married, 1 child—interview n. 1)

Everyone had a story to tell about a lost application, about being excluded from the ranking lists because they had been declared deceased, about a wrongly calculated position or error in the points system, and about a telegram that was never received. But in the will to "resist", or, perhaps, "persist", in the accumulation of stories that have been experienced, felt, told, and modified, they began a learning process. Learning slowly from their mistakes, recognizing people whom they could "trust" (generally relatives who had already got in and were more experienced and who "motivated" them) or friends/colleagues who were in the same boat, as well as the saving grace of "spreading the word", by considering and imitating the profitable choices made by others, and, above all, by simply being "present", they were able to create a strategy that took them from the status of novice to that of "expert" worker, symbolized in the frequently used metaphor of "keeping your eyes open". The awareness which accumulated, thanks to these experiences, was actualized in the contemporaneity of the action, where the subjects were both victims and perpetrators, both subordinates and in opposition, learning to move along a pathway of "plausibility" which allowed them to move from the informal to the formal *frame* without contradiction, as evidenced by the long struggle for stabilization, starting from the anchorage to the finance law, which led to eight months of "occupying" just to obtain a "right", seeing that for once "the law exists" (Male, polyclinic hospital, aged 49, married, 2 children, interview n. 20).

5.4 Cynicism and Nostalgia: Being Unemployed in a Time of Crisis

It is within this framework, therefore, that another segment of our study develops. This part of the research was conducted in Messina between June 2011 and December 2012. Thirty-five in-depth interviews with Italian constructions workers, aged between twenty and fifty, living in popular neighborhoods or slums, were utilized—together with sixteen months of participant observation at the provincial headquarters of a trade union (Fillea-Cgil).

The interest in this category of witnesses arises, as well as to a precise request of the local trade union confederation, from the observation that in the observed context the constructions remain the privileged space of inclusion and reproduction of a large population of natives excluded by the tertiarization processes due to their insufficient education. People were attracted to this particular market because of a relatively low-entry threshold.

The investigation was intended to scrutinize political orientations, survival tactics, and biographies of individuals placed in a "marginal" world made of precarious work, temporary migration, poor housing conditions, oscillation between work and unemployment, and, finally, prison. In sum, the same background of the people we met at the Polyclinic hospital, except for the different perspectives faced by the two "samples". Unlike the previous population, the construction workers who we met in this segment of our study had no chance of being stabilized. Moreover, the years during which this segment of the research took place were the most dramatic for the local construction industry. The crisis had hit hard, and unemployment rates in the sector had skyrocketed. Unlike the hospital workers' case, no "club effect" was available here, unless one considers the trade union itself as one such club. Indeed, the trade union was considered a safety net—a resource to be mobilized against the abuses of an agonizing and exploitative labor market. Yet, as "anyone knew that the unions had reserved quotas", these workers cherished the not so secret hope that, in the future, when public works would have been resumed, union membership would have helped being hired. However, while this assumption was presumably exaggerated, at the moment, for these men,[6] there were no chances other than to face difficulties on each passing day. Depression, sadness, and frustration were, thus, the most common sentiments among workers.

These latter were contacted randomly among members and "sympathizers" who have turned to the door of the federation during the 16-month duration of the research. As most of these workers were resident in those neighborhoods at the center of our interest, we avoided, if not rarely, selective processes based on criteria other than that of age and nationality. When one of the managers suggested we listen to some workers, this occurred because of the history or the prospective personalities of the interviewees, which were considered uncommon. If such a procedure served for the purpose of a comparison internal to the cases, in general, selective processes of this type have been carefully avoided, in order to find, to the possible extent, the exact opposite of the exceptional: normalcy.

Thus, coming to the biographies of the respondents and their view of the world, the first thing that can be observed is that, although we collected individual stories, those heard are collective narratives. That is, these are stories that remark, in first place, the positions of the tellers as "victims" of tort—sometimes one great injustice, sometimes an infinite series—and, secondly, the importance of work for conducting a decent life.

However they are also stories of extraordinary cynicism and "adjustment", particularly in relation to public sphere. Stories that must be read together and in the light of the dependency relationships established by the surrounding powers (authorities, brokers, and market). Narratives of political patronage, certainly, but also individual ways—say, personal policies—designed to navigate within a context of deprivation and to "resist" marginalization.

> T.—To me this politics, these politicians, are only crooked politicians, not real politicians. Because for me the real politics... I can guarantee you one thing ... that for me the real politics was only in the '70s, '80s and '90s. That was the real politics, the one that promised you, the one that was really something for the city. One that actually said: "Yes, now I can't give you a job, but provide me with 50 votes and in two months I'll give you a job". And I can tell you one thing ... I've known people who have a certain age, who are now also retired, who are the people who have benefited from these people ... because at that time there were real politicians. Politicians who had white hair, but that did comply. For me that was a real policy! Beginning in the '80s. I know little of the '70s because I had a certain type of age; But in the '80s, I was there and, say, I met the real politics. There were politicians who ran this city, you won't believe how beautiful was that time. But is it better now that we are dying of hunger? At the time the money arrived in this region. To me, today's politics... I ... I've seen there is no serious politics. You'll have to realize what we have in the House, the Senate ... these people who take 600,000 euros a year to ring a bell: "Please Honourables..." ding ding ... 600,000 euro a year ... these are slaps against poverty! People are starving. Disbelieve that there are people who do not complain. See that there are people who really complain because they no longer have where to get hold of, because, regardless of where they cling, they glide from all places. As I am doing at the moment ... I grip on all places, though I chute; because there is no foothold on which I can cling. But I can also say one thing: for me nowadays there is no serious policy.
> I.—But do you think it was right that to find a job you had to provide a politician with 50 votes?
> T.—No, it wasn't right, but it was something that worked. You know why? Because there was also the politician who gave you coupons for gasoline, the one that gave you 20 pounds of pasta. At the time this was widely accepted. But what he promised you, he really gave. Because he would say that, straight

> in the face: "I can't do it now; however, in two, three months I'll do that". He said that clearly. I can tell you, do you know how many of these people I've known? A person has promised to my cousin a post at the hospital. He told him: "you must give me ten millions lire", it was the 1990s–1995. His mother told the guy: "I'll give the money right away, but the day after my son must swipe the card at the polyclinic"... In the '70s, '80s, things were different ... my father, look, he was a municipal employee; the other brother was an employee at the Region, the other uncle was a technical employee at the office where I've been too for a bit, and the good soul of my grandfather was director at the Satellite building, at the dog-catcher office. I regret that my grandfather died in '74, because if he died in the '90s it was different now. At this time, say, back in the 80's, do you know where I was? I'd be sitting at the city with him, be sure. But my grandfather also had the courtesy to let three children in there, all of them with a job. Never had problems. To me, nowadays there is no longer a real politics. I'll put it in writing. (Turi, aged 48, constructions worker, interview n. 1a)

Albeit with paradoxical traits, these are narratives over spilled sweat and on opportunities and resources unequally distributed; stories on the fatigue to make ends meet and the desire for revenge. But these are also class stories that, among many things, reveal a lack of desire, time, and, above all, the possibility to imagine one's belonging to a collective body. On the contrary, for example, of the stories handed down by Portelli and collected in Terni or Harlan County (Portelli 2011, 2014), what is lacking here is a collective narrative of struggle. More easily, indeed, we find an exaggerated individualism. In these stories, even the union is nothing more than an individual tool available to conduct a series of personal battles.

You might as well say that if many classical historical and sociological studies conducted in trade union circles or among movements were investigations around the ways in which dissatisfaction and exploitation have produced changes, this is instead a study on ordinariness and the reproduction of subordination, despite the presence of political organizations that should prevent exploitation and marginalization. It seems in fact that this local history appears populated by subjects dominated by the circumstances and the relationships. Individuals, therefore, do not live neither in real conflict nor in exasperated relationships with others—since it is precisely by such relations that, as asymmetric and predatory as they can be, personal reproduction depends on.

Albeit prudently, we could also talk about a "new working class" (Ainsley 2018), active in a segmented labor market, characterized by low levels of education, marked individualism, familism, anguish about the future, and even a limited capability of reading the present. But it is still difficult to speak of a "new" working class. The lower classes of today, for most part, in this city are in fact the same as yesterday. They occupy essentially the same neighborhoods and the same positions as their ancestors, according to a social order that has developed and reinforced in the aftermath of the 1908 earthquake and, again, at the end of the Second World War, when the agricultural crisis and the inadequacy of the agrarian reform propelled the internal population of the island to the coastal cities and the housing projects (Finley et al. 1987). The projects and the deteriorated housing stock, again, that in the case of Messina had emerged by the seismic crisis. In this framework, some individuals and households can perhaps experience forms of social mobility and, as we have shown in the previous parts of this book, the subjects are never passive; on the contrary, they are able to act and build paths to overthrow the subalternization, but our focus is rather on the endless cycle of reproduction of such class as a structural effect of the dynamics of reproduction of capital and contemporary capitalism (Wallerstein 2004; Harvey 2010). The same cycle that prevents poor houses to be empty (unlike the high-quality housing stock of the center).

Thus, it appears clear that social mobility seems to have affected the members of this class only marginally. That their education levels remain very low. That their exposure to occupational blackmail in the urban context recalls the semi-feudal rural past and, finally, that their individualism is the result not so much of the "contemporary" (neoliberalism, post-Fordism, etc.) as of a "training" or a "knowledge" that developed over decades and centuries. Over the course of that repressive continuum, in short, which connects the cycles of the socialist peasant and urban struggles in the nineteenth and twentieth century Sicily to the earthquake, Fascism, and post-war Italy (Lombardi Satriani 1979; Scolaro 2008; Ambrosi 2009). Must be this "memory" (a special type of memory; basically, an instinct) of the political defeats, the police beatings, and the betrayals inside and outside of the class that pushed these proletarians to take those cynical and paradoxical attitudes we have seen earlier. This aspect deserves some clarification. Here, we do not want to fall into the trap of cultural determinism that has so often characterized interpretations of many studies regarding Sicily, the mafia and the "Southern Question". We should point out, in fact, that certain ways of crossing the public sphere, and the ethical orientations that lie beneath certain practices arise from a "moral training". That is, from a large collection of frustrations accumulated over the course of their relations with the surrounding environment and the consequent perception of the improbability of any existential breakthroughs—within a context, moreover, of great personal responsibility. It is rather that the ambivalence present in Sicilian culture, that sees life in terms of uncontrollability and decadence, is also associated to a fatalistic attitude, so that the subject remains solidly anchored to the present, almost as if this concrete reality were a reaction to the ineffable uncertainty of the future (it is no coincidence that in the Sicilian language, there is no "future" tense, and actions that take place in this time dimension are always expressed in the present tense). These cultural aspects have been well summed up on a narrative level by many of the most important Sicilian writers: we find them in the family saga described by Giovanni Verga in his cycle of the vanquished, in the image of the Sicilian as a mussel that remains fiercely and solidly attached to the rock because the very moment it detaches, it is inevitably destined to be swept out to sea, like the boat belonging to Padron Toni. It can also be found in the melancholy decadence that surrounds Prince Fabrizio in Tomasi di Lampedusa's "The Leopard", whose existence is imprisoned in a constant existential precariousness, which cannot even be shaken even by the arrival of such extreme political changes as the Unification of Italy because "everything needs to change, so everything can stay the same". In the face of the disaster of impossible change, nothing remains but to take refuge in the present, in what we have, whether it be something material (such as the precious *roba*, the stuff, which Verga mentions in *House by the-Medlar-Tree* and in *Mastro Don Gesualdo*) or immaterial (the heavens, astronomy, the fixed and immutable stars for Prince Fabrizio in the *Leopard*). In one of his works, Pirandello underlines the transience of life in the phrase "Right you are (if you think you are)", showing that even reality is not real, it does not last, and is wrapped in appearance and ambiguity, in a game of veils that corrodes: it may never have existed, it can disappear in the blink of an eye (as in the case of Mastro Don Gesualdo's "stuff"), to be forgotten forever, in a condition in which we are present, but do not exist (as happens to Pirandello's character of *The Late Mattia Pascal*). "Right you are (if you think you are)" expresses a surrender in terms of communication and the difficulty of building truly reciprocal actions because it is not always possible to have absolute faith in others. In the words of one of the most common sayings *amici e vadditi*—which can be translated as "we are friends, but keep looking over your shoulder, because I might betray you".

Within this ambivalent frame, we can see the whole story of Messina: a lot of fuss made over a change, which is described using a lot of big words such as "requalification" and "reconstruction", so that a lot of "hot air" is produced to cover up what everybody already knows, that the change is false to the extent that it is infinite, and will lead nowhere. Once again, here is the idea that "everything needs to change, so everything can stay the same", seen in the reproduction of an order in which the socio-spatial hierarchization of the population remains unchanged. This is why the shacks, the precarious jobs, and the unfairly earned benefits acquired by exploiting the uncertainties of the system, precarious as they are, represent for many the "stuff" to cling on to as hard as possible, even if they are then condemned to stay in an informal space. In a frame such as this, the present is the only place in which survival "tactics" can be put into practice, slowly but surely, to help create a more stable future, like little anchors cast out a bit further each time, in order to try to create a smoother and less troubled line of navigation through the water. Without these tactics, without the possibility of putting them into practice, the future means anxiety, tension, and agitation, and it is lost before it can be earned, as reflected in the stories of these workers.

> T.—Unfortunately, my obsession right now is when I get up in the morning and do not know where to go, that's my obsession. And that's something that brings you the brain 360 degrees … because then you do not know where to turn. Because you know that, day by day, it takes money, you have to eat… You, you can arrange them however, but the children can't always be accommodated. Every time you try to give them something because it's right. For the moment it is so that one must try to adapt. But one can do it relatively, because there are times and moments … there are moments when the brain does not take any more, goes … because I'm already months that I don't work. It disgusts me, it disgusts me and it doesn't take my brain in place. It is not easy, 'cause you know what's that matters in the morning? One gets out of bed and he knows that, in one way or another, he's to go to work, for what it is. But you know, we go. Even under water, what matters at the moment is working… But when there isn't even that, believe me… The evening, I go home and see my children and my wife … and it's as if I don't feel happy. 'Cause, you know, it's a critical time and I don't know how to react. You must try to keep your brain in place because that's what drives you … and if it commands you in the wrong way, you're done! 'Cause you've limits in life… When you arrive to that wall that you can't cross, you have to stop. But if you want to cross that wall, you take a bat in the hand.

That's why I say that the brain turns and turns, sometimes good and sometimes bad. Sincerely, certain evenings I am ashamed to look at my children in the eyes, and to my wife. But not that I'm ashamed… It's something that I carry within me because right now I'm not working, but … I try to be always in some kind of harmony…
I.—It's not your fault…
T.—It's not my fault, because otherwise I'd be a piece of shit. It's a general fault that we are all suffering, and there are people who are worse than me. But, at this time, that's the thing. But I have the humility to say that right now I am ashamed. My children don't say anything because they know that I've always worked. Though I think that I am ashamed. Do you understand? (Turi, aged 48, construction worker, interview n. 1a)

A.—For now that I'm unemployed, my typical day is not to sleep at night because I don't work during the day. I'm not tired at night… I do 3, 4 am, and I don't sleep; I get to sleep at around 5 am. I fall asleep and in the morning. I sleep properly then… And, then, you do noon, 1 pm… I get up, eat, and then in the afternoon I don't know what to do, because I can't use the car. It still gets a charge, the fuel has a cost…
I.—What car do you have?
The Alfa Romeo147. It costs … so for now, it's summer and I did not go even to the beach. Today for example, I got up early because I had to talk to the accountant. And afterwards, I stayed home … computer, television, and then I had lunch at 1. Then I watched TV… Afterwards, at around 4 pm, I took a nap and now I'm here. After we finish, I should go running. I like running. Yesterday I couldn't go because I have come of leg pain as I don't have shoes suitable for running. And since I'm not working, I can't go to get hold of this new shoes… Because since they're old, they are no longer good, they are consumed. If I were working, I'd buy them immediately, immediately, because they are something I use. Last night, it looked like I had two icicles … because it hurts and I didn't go because otherwise today it'd be even worse…. (Antonio, aged 20, constructions worker)

That unemployment and insecurity have important repercussions on the emotional (and public) life of individuals is not new, as evidenced by an extensive review of studies on the subject conducted by Kalleberg (2009). The reason why it is important to dwell on the issue is that it is only by subtracting the reflection to the rule of pietism and to the "psychologisms" that we can understand how the subaltern order reproduces, and also opposes, the cultural essentialisms.

It would be to say that subordination is not merely a state of subjection. Forcing Berger and Luckmann (1966), we could argue that subordination consists of a state of subjection powered by the economic structure of a place and by an overarching ideological apparatus, the effect of which is to generate shared visions for those in contact with such set of formal and informal institutions—visions based on the similarity of the experienced feelings.

"Cynical" orientations and political practices—some would say clientelistic, mafioso, and so on—are thus the result of a life organization intended to produce emotions and, subsequently, courses of action, strategies, tactics, and justifications. Therefore, analyses limited to record simply the frequency with which actions and speeches repeat (e.g., the longing of many respondents for the time when it was enough to procure votes to get a job) are likely to result in the explanatory terms of a failure of the political civilization, or in other equally culturalist and moral terms.

But the struggle of certain respondents not to succumb to "certain thoughts" (such as "to take a bat in the hand"; see Toni's excerpt above) and repeat gestures that in the past have led them to the prison, as well as the widespread nostalgia for the years of political patronage, bespeak only of a search aimed to counter a forced subordination.

Above all, the stories of this inner conflict reveal an attitude that, despite appearances, is fundamentally political. Merton (1968, p. 221) observed that under capitalism, individuals develop a tendency to ascribe to themselves, rather than to the "system", the responsibility for the failures. If this is true, we can consider the average condition of the citizens active in this regime as the manifestation par excellence of subjection and impoliticness. None of this, however, is found in the words of witnesses who appear variously tormented but who assign to the "system" the responsibility for their own marginal status.

In this sense, these respondents—as well as many of those who share their destiny and experiences—are still political individuals, despite being objectively context-dependent. And despite the fact that they can, in an attempt to free themselves through unorthodox and informal practices, participate in extending the number of individuals subjected to unethical powers—thus reproducing the oppression.

5.5 Subaltern Politics

In the heard narratives, there are some recurring themes: for example, the sleepless nights, and a certain way to referring to divine and equality. A worker among all, Dino, in his late forties, answers a question on the issues of political participation by bringing up a nocturnal intimate scene:

> Interviewer—You think you are an individual, an island in its own right, or do you think you belong to a group of people sharing work-related problems, linked to exploitation? You know what I mean? Do you imagine yourself as a single person who has to face a number of problems: the employers who do not pay, the bank asking you money, or you think you are part of a group of people who are oppressed by these people?
> D.—But I do not know… Look … I only know one thing … that sometimes, not working … not working, the brain … fucks. I have not slept for three nights.
> I.—Why?
> D.—I do not know, perhaps because … because a person thinks, he thinks that he has a family to… Even if you find the work, you also find… Say, you find that one who takes advantage of you, who doesn't give you what he must give you … and then … really, really, are … My wife told me: "Take sleep drops" … but such drops … I need a valium bucket to sleep … and then the brain, the brain … fuck, it nags you, it pesters you … you feel … you feel … do you know what once it's happened to me? I'll tell you … I went to the health insurance fund, fuck, for a medical examination; I wasn't working that time … fuck, look what happened to me… I went to book a physical with Dr… I don't remember her name … fuck, it just so happens I was put there so sitting, and I heard people talking to each other… You know that I got so sorry that I started to cry? The doctor then called me and asked me "why are you crying?" Look, certain times… I cried because they were telling their troubles … well, fuck, there are people who are worse than me… I'm really sorry… We had to be all rich, or poor… The Lord has done many things, many beautiful things, but on one thing was wrong, one thing… just one thing he did wrong: two hands and ten fingers. Instead he had to give us a finger and ten hands. (Dino, aged 48, construction worker, interview 6a)

Insomnia, the jokes exchanged with his wife in bed, the reference to valium to mitigate the vortex of thoughts, they all suggest that his primary reference sphere is the individual and private one, represented by the family. But soon the story turns into an empathic narrative that relates to the outside. Dino, in fact, said he heard the speeches of other people waiting for a medical examination and burst into tears. Despite its isolation, then, he tells us that he is still capable of suffering for others because he pities them or, perhaps, because by hearing them he experiences a catharsis ("then there are people who are worse") which, as is typical of this process, makes him feel momentarily relieved. But that is not all. Dino also says that he is bothered by inequality. This is expressed through the contrast between rich and poor, as well as through a series of images and religious references to the universe. Considering the locality and the position from where he speaks, in doing so, he shows a kind of continuity with the typical forms of popular religiousness and politics of the rural South (Hobsbawm 1959). Although it does not refer to a socialist Jesus—a typical motif of the "conflictualist" popular religiosity—Dino still leads the theme of equality on the plan of nature and, therefore, of the divine. If we are not exactly in the territory of the religious imagery of struggle, however, we move inside a related field of representations and cultural reordering of the world.

For the purposes of a discussion about new popular classes and their new characters, we could at this point ask ourselves if there is any continuity between a certain way of representing social order, inequality, and even to oppose class divides. And about the theme of continuity, we might as well assume that the questions asked by Dino do not differ much from those of the southern, rural workers of the nineteenth and the early twentieth centuries. We have not insisted much on the characteristics of labor in the late nineteenth-century post-feudal Sicily, but the main features were the presence of an abusing landowner, a number of median figures, and a large base of daily workers subjected to the whims, the abuses, the needs, and the economic and political interests of the "command chain" above them. At a certain point, this societal organization had moved to the city, where it transplanted practices and relationships in completely new environment (Arlacchi 2007; Lupo 2011). In given circumstances, the big firms, subcontractors, gang-masters, and daily and semi-daily workers are the contemporary "equivalents" of those rural figures, especially in a system characterized by high rates of informality. We have already underlined (see Chap. 3) how the peasants were not simply salaried workers, but mixed figures, often owning small areas of land which they cultivated for themselves, or carrying out manufacturing activities, also from home (such as small artisanal activities or the transformation of raw materials; in the case of Messina, e.g., the production of silk as a cottage industry). The role of the peasant as a "mixed figure" can be seen in the context of the wider productive and reproductive strategies of the rural family, which was at the center of economic organization: formal and informal, production for the market, for domestic consumption, or for the economics of reciprocity. These were all part of a complex system of checks and balances that allowed the peasant and family members to resist impoverishment and existential precarity. As Signorelli (1983) points out, the peasant family was a veritable safety net. It was only when this balance was altered and the "mixed-figure" peasant became an increasingly proletarized salaried worker, that these phenomena of progressive impoverishment worsened and could only be resolved by mass emigration.

In this respect, there are evident elements of continuity and differentiation. On the one hand, the centrality of the family emerges once again as fundamental to the development of survival tactics and social mobility. However, these remain precarious and unstable and, much more so than in the past, linked to mechanisms of opportunistic exploitation (such as improper access to welfare benefits) which, in the long term, erode the collective fabric and create forms of tragedy of the commons.

On the other hand, what emerges is a gradual process of proletarianization and the commodification of the economy, so that it is increasingly difficult to create subsistence without the market or without salaried work (Polanyi 1944). This has meant exposing workers to those hidden, informal, and clientelistic forms that lurk behind markets and increase costs, and which, by creating murky relationships, make it easier to exploit the weak and widen profit margins. Contrary to current opinion, these are not the characteristics of a backward society that is destined to disappear, but market operating procedures that always work in a monopolistic or oligopolistic way (Braudel 1985; Wallerstein 2004). It seems likely to agree upon the notation that, basically, today's levels of exploitation in low capital intensity markets such as the one under consideration—as well as the exposure to occupational blackmail, and the dependency of workers by contractors—appear as a substantial transposition in the urban environment of the nineteenth-century southern, rural insecurity described, among others, by Hobsbawm (1959). The same insecurity and "dependency" that, over the course of this book, we have seen in action in the subaltern spaces of the city, before and after the earthquake. The relationships and the work modes appear, in fact, functionally overlapping, characterized as they are by day labor, low wages, irregularities, and impossibility to assert rights.

If this is the structural framework, two particular aspects of the discussed problem refer to the ways of opposing this situation and the political forms of dissent. Here, one can see some substantial differences. The most obvious is that, as far as the southern country has not always been a place of rebellion, in the past bloody riots and self-organization characterized the country life far more than they do now (Hobsbawm 1959). Thus, returning to the "collective character" embodied by Dino, and the workers of polyclinic hospital heard before him, we come up against some kind of contradiction. We have seen that these people are not, even if only at a gut level, less "political" of their peasant and proletarian counterparts of the past century or so. They feel, in fact, the "weight of the world", to put it in Bourdieu's words (1999). Is it really possible, then, that demands and aspirations for redistribution and equality are kept on a plan of inaction?

A look at a problem that was particularly vexing in the period when we met Dino allows to advance the hypothesis that he, like many others, is active in reducing the gap, but that his "political commitment" does not manifest itself in places and forms normally considered as such. The problem, in fact, is to recognize the political nature of these practices apparently very distant from the political. The same practices, moreover, that the local common discourse interprets in terms of "(sub-)culture" and "missed civilization".

More specifically, Dino had for some time a problem with a finance company. After years of timely payments, he found himself unable to honor its debt. Thus, the typical drawing of phone calls and threats from collection agencies responsible for the recovery of due figures began.

> D.—I haven't worked since June … the fact that I explained to you before, and I am hoping that… my mother, my mother-in-law can help…
> I.—Little jobs on your own? Didn't you get any?
> D.—Well, when I get one… But only small things… whatever…
> I.—What other jobs have you done this time?
> D.—Well, a friend calls me to make a move, or moving furniture from offices to offices, or cleaning offices a few weeks, ten days… or washing stairs… Let's say, rightly enough that it's a clean job with… this thing of finance company [laughs].
> I.—Tell me about it…
> D.—Oh, then I, for ten years in a row, I have always paid this company. I have always paid.
> I.—But the first time that you entered in this debt circuit, why was that?

D.—So, I have made a loan and I bought a car. Then I borrowed 15,000 euros, about ten, nine years ago … then I've paid, I always paid on time. I paid 850 euro a month … then, as soon as I paid back, they gave me further money… But the bottom line is this one: I've always paid the company. But now, since June I no longer work … that is, how can I continue to pay? I can't pay them anymore … Not that you don't want to… I mean, you'll do it… because it's right, because they helped me. But now they've sent this telegram: "pay 3,000 euros!", and the 3,000, where are they? For ten years, you have always paid on time… Still, a punctuality that they themselves said, when I went there… I need 3,000, 4,000 euros, that's it… They not even made me talk … But not that I ate this money. They were always needed for the home, for the children, this and that. If all of a sudden I find myself up against the wall and I can't pay any more these 800 euros, between mine and those of my wife… I mean, you can't put me up against the wall. That is, they can tell me "Mr. Cucinotta, I don't mind about what you do. You have to pay this money", but put me in the position… We've sent a fax. Currently, I don't work, I can give you 100 euro a month. Then, when I'll work again, I'll continue to pay normally.
I.—What are you going to do now?
D.—With the company? Ah, now we sent a fax and asked this thing, if for the moment… But then, what they told me… Even earlier, I got a call from Florence, saying: "Oh well, given that you don't work and that your wife works part-time, put together yours and your wife's money and do 500 euro per month". Excuse me, if my wife takes 400, 420, can you tell me how do I pay 500 euros a month? I have stuff to do with this budget… You'll tell me! [laughs].
I.—I can't really tell … [laughs].
D.—But we can't do anything… Also… It's not that I want to scrub; because if I wanted to scrub, I wouldn't have paid 800 euro a month for ten years. Do you understand what's the point? I mean, I was always on time because they favored me, and I am punctual … but if I don't work, I can't go and steal… and if I'm going to steal, I do it for my family, not for a company… oh, I am honest…
I.—You own the house, right?
D.—Yes, I'm in a shack! A cottage… it's a shack. The company claims that I lied. I lied? Then, when I went there, they asked: "The house is yours?" Yes, it's mine … and then they wrote in the contract "home owner". But, sorry, I said faithfully "The house is mine". Certainly it's not yours, it's mine… Is that right? In this house lived my brother in law, and then I went to live myself in this house, right? But this house is not in the Land Registry, and not even in the mind of the Lord. I mean, I mean, the company asked "Is the house yours?" Yup! "Fine, Mr. Cucinotta. Bring me the contract!"

Is it right? And I would've answered: "I have no contract", because I'm in a small, illegal house. A house of 29 square meters, unfortunately... not even 30 meters... What should I've said to the company? Come here, so that you can find the mice walking on the roof?! No, really, I swear to my children... The thing that bothers me more are these claims. That they say "Pay! Pay by tomorrow! Pay 3,000 euros". Pay... Am I your brother?! Not that I won't give them to you... but what do you know about my condition? [...] What I have to say to that company? That for now we are chasing hunger with the motorcycle? And they say: "pay 3,000 euros!" But put me in the position... Meanwhile, for now take these hundred euro a month, that it's possible that you lose those as well ... No, the company... Really, apart from the jokes, you know ... it became a thing... I'll tell you about this girl who called... So I had a phone call from this girl of debt collection agency. "Mr. Cucinotta, good morning!" Good morning! This is because my wife hasn't paid for a month... "Mr. Cucinotta hello, there is a missing payment". And I replied: "Look, lady, miss, I am not working currently, I can't pay ..." "No, you have to pay." "But I can't... if I was a woman, say, I would do something and I would pay, then..." "Okay" she goes, "but you can do the gigolo"... Shit, for the love of my children, for real, she got to up that point where she said to me that I can do the gigolo thing...! "Let me do it first with you and your mother, the gigolo!" Fuck! Then, another debt collector... Since I have insulted them, my wife told me "it's useless for you to treat them this way. Hung up the phone"... Another time, more debts... Each one more infamous than the one before, these people. "Good morning Mr. Cucinotta, you've to pay!"... "I can't pay"... And I added that I would have turned to the lawyer. "But you can also call the President of the Republic", she said. "Look", I now tell her in good Italian: "Look, scumbag, but who are you?" And I hung up the phone. My wife said: "sometimes you'll find yourself in a bit of jam". Yes, but if you treat me like shit, how could I be any different? (Dino, aged 48, constructions worker, interview 6a)

Switched off the recorder and questioned again on the matter, Dino explained that the deepest reason he had to submit himself to contract loans and pay his debts for years, and punctually, was tied to a dream: he wanted to end his life of debtor with a big scam. The biggest scam that he could think to accomplish. In fact he sought to pay off its current debt as soon as possible, in order to qualify as a reliable customer and still be young enough to have potential access to further credit. What he wanted was €50,000: money to be taken and given to his children, his wife, and his closest relatives, and not to return. He missed very little to pay off its current debt, and thus concluding the first phase of the plan. But the crisis had come and his secret dream was shattered, leaving room for the persecution of the finance company.

How to read these words and, therefore, this project? The story of Dino is basically that of hundreds of thousands of people, trapped for various reasons in the vise of consumer credit. His is the story of a poor chased by another poor—these latter employed in call centers in charge of debt collection, the owners of which are, at least in the South, able to pay even less than €4 per hour.

If only for this, the story sheds light on the daily lives of those who, today, survive poorly in both the classic places of exploitation as are the construction sites, and in the modern "traps" of advanced services for the enterprise. In other words, it shows a new way of interpreting the old dialectic between "cops and robbers". The old opposition within the proletariat or the impoverished middle classes that pushed the poor to become either a criminal or a policeman, sees now the expansion of the range of possibilities, by offering crafts such as call-center operator (Woodcock 2017). This operator specializes in tracking subjects that are insolvent and often not readily available. He (or she, since the operator is more often a woman) enters a phone number, or digs into databases, and cross data with the professionalism that comes from the desperation of a "commission" work that, usually, ensures only skinny compensation. So, the new proletarian—beyond his old enemy, the cop—has a new Cerberus to fear: the finance company and its "bloodhound", the call-center operator.

Dino's one, therefore, like the history of Messina and its "underclass", is a story at the same time old and new. It is a story of our time, which takes place in the age of finance companies, the artificial boost to consumption and debt, mobile, Internet, and databases that follow the individuals, but it is also a story of the past, culminating in the older and classic ways: with the poor being chased by the bloodhound on duty (a tax collector, a royal policeman, a tax collector, a call-center operator).

Of course, the poor in question is also a crook. It is he who intends to cheat a finance company to give a bit of money—not much, but still a substantial figure, especially if you are a penniless forced to work as a slave for a few hundred euros per month—to his loved ones.

Dino's plan is perhaps unrealistic and unlikely to succeed. But it certainly expresses a will of appropriation which is also reactive. To employ the old jargon of the Italian radical left, he, instinctively, resuscitates the idea of a "comontista" gesture—that is, an action turned toward the immediate, guided by the idea, individualistic and yet "revolutionary", that it is enough to transform the everyday life to build a better world (at least for the person who has the courage to act) (Vallauri 1976, pp. 109–110). The idea, in short, that sometimes social justice implies lawlessness and incivility.

5.6 Final Remarks

Despite the striking differences between economic and political regimes, technologies, and forms of work organization guarantee institutions and security bodies, it is possible to believe that the current characteristics of the unskilled labor market are, in this city of the south, an accurate transposition of a social organization that was born and formed in the countryside of the nineteenth century, into an urban environment—based on daily labor, exploitation, no rights, and no regular salary. Thus, the fundamental lesson that we can draw from the cases explored here is that capitalism is fundamentally a less disruptive and innovative force than we imagine. At least in some cases, it appears to feed on past relationships and produces changes that are merely trivial.

The study of these Sicilian precarious workers in a post-earthquake city shows, in short, the persistence of the social relations and structures within the labor market that challenge those "modernist" representations which focus, for example, on the dramatic changes of the national economies, new jobs, start-ups, and technological changes. In fact, in addition to specific geographic areas, especially if they are metropolitan, and to a limited number of sectors dramatically affected by these changes, a great many other realities are either excluded or only marginally hit from these processes. Or, like Messina, encompassed in permanent cycles of crisis and transition.

Our analysis, thus, challenges both enthusiastic and critical interpretations of the present that focus on change. Although we do not deny that these social and technical changes exist and alter the worlds of life, we think that the ideas of continuity and persistence are also important. In particular, we attempted to show how dramatic changes of the past—those changes that compose our present, our indirect memories, and our perceptions as human beings belonging to a world of life—were at the same time producing novelties and reproducing the past. The disaster, the new city, the character of the bourgeoisie that emerged at the turn of the century, the raising of welfare, the transformation of the land uses, and the shift from the rural to the urban (in terms of life styles, jobs and housing) were, in the case of Messina, impressive changes. Nevertheless, all these shifts over the course of a century have not been able to alter the "substance" of the relationships among players, and also the nature of the relations between areas, that is, the north and the south of Italy, Messina and decisional centers. Rather than focusing on the changes, our interest was for the continuity of multi-level asymmetries. For that stability of the entailed relationships that, in our view, help explain the reality and also to interpret the course of history in a different manner. In our perspective, the Italian *Mezzogiorno,* that appears so "essentialized", "orientalized" (Schneider 1998), and "allochronic" (Fabian 1983)—the latter concept being well summed up by the ideas of "delay" and "double speed", deeply rooted in the public discourse, as is recognized by anyone who shares the national "intimacy"—ceases to be the site of a congenital material and moral backwardness, and becomes a prefigurative space of modernity. Still an allochrony, it is true, which, however, just like the history of the colonies (Stoler and Cooper 1997), makes tangible the artificial nature of some territorial hierarchies, as well as the weakness of many related hegemonic narratives, centered on the opposition between development and underdevelopment. Below the surface of the transformations and the modernization lies, in fact, the essence of real social relations, which are characterized, in general, and increasingly so over the past decades, by the instability of life and career paths, by the temporary nature of labor performance, by low wages, insufficient welfare, and intergenerational dependence for the purposes of sustenance, and, finally, by insulation, police repression, inequality, and polarization (Bauman 2011). Which of these keywords, then, should sound new when referring to Messina and the south of Italy?

Notes

1. Following Bourdieu (1984) from an epistemological point of view, it has been possible to take advantage of our proximity to the context being analyzed, blending the data collected with the insight gleaned from closeness to and empathy with the research subjects. In our view, the interview appears to be an "active" process (Holstein and Gubrium 1995; Fontana and Frey 2008), aimed at creating a relationship between the parties involved in the dialogue that is based on trust, at reconstructing their experiences, and at stimulating a rich and full narrative. The face-to-face interviews were conducted mostly in the subjects' own homes, sometimes together with a

spouse, and aimed at retracing their family and life history. Other interviews, specifically those with the construction workers, were conducted in the trade union's headquarters.

2. With Perlman (1976), we believe that "marginality is a myth", and that the urban poor and disenfranchised, in general, are, on the contrary, fully integrated into society, but in such a way as to be economically exploited, politically repressed, socially stigmatized, and culturally excluded. See also: Bayat (2000, p. 539), Alsayyad (2004, p. 9).
3. Low-skilled service jobs are often manual activities, which are labor-intensive, but they require low production volumes. The disparity between industrial productivity, which is in continual growth (thanks to technological innovation), and that of services, which remains stagnant, creates in entirely de-regulated markets (such as the informal one) a situation where only low-level salaries are paid for these activities, and which is mirrored in an increase in social inequality (Esping-Andersen 1999).
4. For similar accounts, see: Mukhija and Loukaitou-Sideris (2014), Graaf and Ha (2015).
5. In reality, Law 56/87 only allowed the civil service to employ low-level personnel directly from the ranking lists compiled by the employment office, on the basis of certain prerequisites (in particular length of unemployment). In this way, it was possible to bypass the public written examinations, which are necessary for higher-level jobs, and to establish a mechanism whereby personnel are directly "called" by the employment office if they have the necessary prerequisites. It is no coincidence that, for the precarious workers who were interviewed, the expression "called" is synonymous for work in the public sector.
6. Unlike the hospital, this sector, as well as the branch of the union that deals with workers, was composed of men only. The few women, who were occasionally seen in the room, were the wives and the daughters of workers who were sick or had had accidents at work, and who needed documents.

References

Ainsley, C. (2018). *The New Working Class. How to Win Hearts, Minds and Votes.* Bristol: Policy Press.

Alsayyad, N. (2004). Urban Informality as a New Way of Life. In A. Roy & N. Alsayyad (Eds.), *Urban Informality. Transnational Perspectives from the Middle East, Latin America and South Asia* (pp. 7–30). New York: Lexington Books.

Ambrosi, L. (2009). *La rivolta di Reggio. Storia di territori, violenza e populismo nel 1970.* Soveria Mannelli: Rubbettino.

Agarwala, R. (2013). *Informal Labor, Formal Politics, and Dignified Discontent in India.* Cambridge: Cambridge University Press.

Arlacchi, P. (2007). *La mafia imprenditrice. Dalla Calabria al centro dell'inferno*. Milano: Il saggiatore.

Bagnasco, A. (Ed.). (1986). *L'altra metà dell'economia: la ricerca internazionale sull'economia informale*. Napoli: Liguori.

Bauman, Z. (2011). *Collateral Damage. Social Inequality in a Global Age*. Cambridge: Polity.

Bayat, A. (2000). From 'Dangerous Classes' to 'Quiet Rebels': Politics of the Urban Subaltern in the Global South. *International Sociology, 15*(3), 533–557.

Benton, L. (1990). *Invisible Factories: The Informal Economy and Industrial Development in Spain*. Albany, NY: State University of New York Press.

Berger, P. L., & Luckmann, T. (1966). *The Social Construction of Reality*. Garden City: Doubleday.

Boels, D. (2016). *The Informal Economy Seasonal Work, Street Selling and Sex Work*. New York: Palgrave Macmillan.

Bourdieu, P. (1984). *Homo Academicus*. Paris: Les Editions de Minuit.

Bourdieu, P. (1998). The Economy of Symbolic Goods. In P. Bourdieu (Ed.), *Practical Reason: On the Theory of Action* (pp. 92–123). Stanford: Stanford University Press.

Bourdieu, P. (1999). *The Weight of the World. Social Suffering in Contemporary Society*. Stanford: Stanford University Press.

Braudel, F. (1985). *La dynamique du capitalisme*. Paris: Arthaud.

Cassese, S. (1993). Hypotheses on the Italian Administrative System. *West European Politics, 16*(3), 316–328.

Castel, R. (2017). *From Manual Workers to Wage Laborers: Transformation of the Social Question*. London and New York: Routledge.

Cerase, F. P. (1990). *Un'amministrazione bloccata. Pubblica amministrazione e società nell'Italia di oggi*. Milano: FrancoAngeli.

Cerase, F. P., & Farinella, D. (2009). Public Service Motivation: How Does It Relate to Management Reforms and Changes in the Working Situation of Public Organizations? A Case Study of the Italian Revenue Agency. *Public Policy and Administration, 24*, 281–308.

Coletto, D. (2010). *The Informal Economy and Employment in Brazil: Latin America, Modernization, and Social Changes*. New York: Palgrave Macmillan.

De Certeau, M. (1984). *The Practice of Everyday Life*. Berkley; Los Angeles; and London: University of California Press.

Esping-Andersen, G. (1999). *Social Foundations of Postindustrial Economies*. Oxford and New York: Oxford University Press.

Fabian, J. (1983). *Time and the Other. How Anthropology Makes Its Object*. New York: Columbia University Press.

Farinella, D. (2013). Getting by in a Postfordian Age: Survival Strategies and Temporary Workers in the Sicilian Public Sector. In P. Saitta, J. Shapland, & A. Verhage (Eds.), *Getting by or Getting Rich? The Formal, Informal and Criminal Economy in a Globalised World* (pp. 95–115). The Hague: Eleven Publishers.

Farinella, D., & Irrera, O. (2014). Eterotopie della resistenza e classi subalterne. Infrapolitica e mobilitazione per il lavoro in un'azienda sanitaria del Mezzogiorno. *Etnografia e Ricerca qualitativa, 2*, 195–218.

Ferrera, M. (1996). The Southern Model of Welfare in Social Europe. *Journal of European Social Policy, 6*(1), 17–37.

Finley, M. I., Mack Smith, D., & Duggan, C. J. H. (1987). *A History of Sicily*. New York: Elisabeth Sifton Books.

Fontana, A., & Frey, J. H. (2008). The Interview: From Neutral Stance to Political Involvement. In N. K. Denzin & Y. S. Lincoln (Eds.), *Collecting and Interpreting Qualitative Materials* (pp. 115–159). Thousand Oaks, CA: Sage.

Geertz, C. (1983). *Local Knowledge*. New York: Basic Books.

Ginatempo, N. (1976). *La città del Sud. Territorio e classi sociali*. Milano: Mazzotta.

Graaf, K., & Ha, N. (Eds.). (2015). *Street Vending in the Neoliberal City: A Global Perspective on the Practices and Policies of a Marginalized Economy*. Oxford and New York: Berghahn Books.

Guha-Khasnobis, B., Kanbur, R., & Ostrom, E. (2006). *Linking the Formal and Informal Economy: Concepts and Policies*. Oxford: Oxford University Press.

Harvey, D. (2010). *The Enigma of Capital and the Crises of Capitalism*. Oxford: Oxford University Press.

Hobsbawm, E. J. (1959). *Primitive Rebels*. New York: Norton.

Holstein, J. A., & Gubrium, J. F. (1995). *The Active Interview*. Thousand Oaks, CA: Sage.

Kalleberg, A. L. (2009). Precarious Work, Insecure Workers: Employment Relations in Transition. *American Sociological Review, 74*, 1–22.

Leonard, M. (1998). *Invisible Work, Invisible Workers: The Informal Economy in Europe and the US*. New York: Palgrave Macmillan.

Leonard, M. (2000). Coping Strategies in Developed and Developing Societies: The Workings of the Informal Economy. *Journal of International Development, 12*(8), 1069–1085.

Lombardi Satriani, L. M. (1979). *Rivolta e strumentalizzazione. Il caso di Reggio Calabria*. Milano: Franco Angeli.

Lomnitiz, L. (1988). Informal Exchange Networks in Formal System: A Theoretical Model. *American Anthropologist, 90*, 42–55.

Lupo, S. (2011). *History of the Mafia*. New York: Columbia University Press.

Merton, R. K. (1968). *Social Theory and Social Structure*. New York: The Free Press.

Mingione, E. (1983). Informalization, Restructuring and the Survival Strategies of the Working Class. *International Journal of Urban and Regional Research, 7*(3), 311–339.

Mingione, E. (1985). Social Reproduction of the Surplus Labour Force: The Case of Southern Italy. In N. Redclift & E. Mingione (Eds.), *Beyond Employment* (pp. 14–54). Oxford: Blackwell.

Mukhija, V., & Loukaitou-Sideris, A. (Eds.). (2014). *The Informal American City. Beyond Taco Trucks and Day Labor*. Cambridge, MA: Mit Press.

Paci, M. (1989). *Pubblico e privato nei moderni sistemi di welfare*. Napoli: Liguori.

Perlman, J. (1976). *The Myth of Marginality. Urban Poverty and Politics in Rio de Janeiro*. Berkeley, CA: University of California Press.

Pieretti, G. (1991). I nuovi volti della povertà urbana. In M. Pellegrino & V. Verzieri (Eds.), *Né tetto né legge* (pp. 31–42). Torino: Ed. Gruppo Abele.

Polanyi, K. (1944). *The Great Transformation*. New York: Farrar & Rinehart.

Portelli, A. (2011). *They Say in Harlan County. An Oral History*. Oxford and New York: Oxford University Press.

Portelli, A. (2014). *Biography of an Industrial Town. Terni, Italy, 1831–2014*. London and New York: Palgrave Macmillan.

Portes, A., Castells, M., & Benton, L. (Eds.). (1989). *The Informal Economy: Studies in Advanced and Less Developed Countries*. Baltimore: Johns Hopkins University Press.

Portes, A., & Haller, W. (2005). The Informal Economy. In N. Smelser & R. Swedberg (Eds.), *Handbook of Economic Sociology* (2nd ed., pp. 403–425). New York: Russell sage Foundation.

Saitta, P., Shapland, J., & Verhage, A. (Eds.). (2013). *Getting by or Getting Rich? The Formal, Informal and Criminal Economy in a Globalised World*. The Hague: Eleven Publishers.

Schneider, J. (Ed.). (1998). *Italy's 'Southern Question'. Orientalism in One Country*. New York: Berg.

Scolaro, G. (2008). *Il movimento antimafia siciliano. Dai Fasci dei lavoratori all'omicidio di Carmelo Battaglia*. Catania: Terrelibere.

Signorelli, A. (1983). *Chi può e chi aspetta. Giovani e clientelismo in un'area interna del Mezzogiorno*. Napoli: Liguori.

Stoler, A. L., & Cooper, F. (1997). Between Metropole and Colony. Rethinking a Research Agenda. In F. Cooper & A. L. Stoler (Eds.), *Transitions of Empire. Colonial Cultures in a Bourgeois World* (pp. 1–56). Berkeley: University of California Press.

Trigilia, C. (1992). *Sviluppo senza autonomia. Gli effetti perversi delle politiche nel Mezzogiorno*. Bologna: Il Mulino.

Vallauri, C. (1976). *I gruppi extraparlamentari di sinistra*. Rome: Bulzoni.

Wallerstein, I. (2004). *World-System Analysis. An Introduction*. Durham, NC: Duke University Press.

Williams, C. C., & Windebank, J. (2003). *Poverty and the Third Way*. London and New York: Routledge.

Woodcock, J. (2017). *Working the Phones. Control and Resistance in Call Centres*. London: Pluto Press.

CHAPTER 6

Formal and Informal Housing in Today's City

6.1 The Reproduction of the Shanties

This chapter aims to underline how the construction of urban public space in the city of Messina follows a line of intervention that, from post-earthquake reconstruction onwards, relegated marginalized subjects to specific areas of the city, housing them in specific types of accommodation, in order to make the most of ground rent and favor speculative building by large landowners—also in order to build pools of intermediary patronage linked to vote trading.

Today, the city of Messina is an urban conglomeration that has expanded longitudinally from north to south over a distance of about sixty kilometers, gradually spreading up into the hills that lie behind the entire city, where large agglomerates of buildings have been constructed. As seen in Chap. 4, the expansion of the city happened according to spatial principles that also express social stratification. The city center and the panoramic northern area were destined for private and residential buildings for the middle and upper-middle classes, while the more working-class popular areas were situated in the peripheral and external areas (the southern part of the city and the more internal and hidden areas of the center and north).[1] These large portions of territory, where some of the post-earthquake shacks were initially located, were destined to become working-class areas and included sections of public housing (which have expanded over the years) which have blended with the urban stratification created by

D. Farinella, P. Saitta, *The Endless Reconstruction and Modern Disasters*,
https://doi.org/10.1007/978-3-030-19361-4_6

the various periods of slum clearance and slum repopulation, to create a chaotic landscape. There are different types of housing here (which reflect the architectural and urbanistic trends of the periods in which they were built), interspersed with "fondi", or groups of shacks and slums, and "casette ultrapopolari", as well as shacks, small houses, and illegal self-builds, in either groups or singularly, built onto the walls of council flats, leaning up against street walls, or even in the courtyards or parking lots of public housing.

The "rehabilitation" is an unending project, which proceeds slowly and erratically. Despite the intense debate and the various appeals to national and regional parliaments from the end of the Second World War, it wasn't until 1978 that the new regulatory plan Tekné was approved, promptly identifying three areas for rehabilitation, although no specific information was given as to priorities. Just twelve years later, with the Regional Law n. 10/1990, the Sicilian Region launched a massive financial project of 500 billion lire for the rehabilitation of Messina. This law required the Municipality to carry out a census of the family units living in the areas destined for regeneration and to come up with a detailed plan for each area to be included. This bureaucratic process took a further twelve years, and only in 2002 did the Census of the family units residing in the area finally get off the ground, and also the detailed plan for each of the seven areas to be regenerated (Regione Sicilia 2012; Comune di Messina 2012). During those years, the Autonomous Institute for Public Housing (IACP) continued building public housing and clearing some of the "fondi", but most of the money that had been formally allocated either was not spent or did not arrive. Of the 500 billion lire stipulated in the Regional Law n. 10/1990, only 150 billion lire was spent, while the rest of the money seems never to have arrived. The subsequent Regional Law n. 4/2002, designated a smaller budget, of €70 million. What is apparent is the local public administration's inefficiency in managing the various phases of the recovery plan, from the administration of the funds allocated for the housing construction (that are often so late that the buildings are no longer viable) to the construction and final delivery of new housing, to the demolition of the shacks remaining in the regenerated areas, and the disposal of the resulting detritus, including the safe disposal of asbestos (which was used as roofing material for many of the shacks during the 1970s and 1980s). One only has to remember that between 2002 and 2018, only 584 houses were allocated to the shack-dwellers (Comune di Messina 2018). Over the years, many newspaper articles have been written about the late or non-existent delivery of housing, citing a whole host of different reasons. Perhaps this was due to work approved on paper, financed, but never executed (Tumino 2016; Lo Verso 2015), or work that was left unfinished by a contractor being investigated for mafia connections (D'Amico 2012). Sometimes it was because groups of shack-dwellers ineligible for allocation illegally occupied the housing (Ipsale 2015), or perhaps the houses had not been certified as habitable, or the shack-dwellers had paradoxically decided that they would only leave the shacks, to transfer to the new housing, during clearance, so as to avoid new reoccupations. Again, a clearance not carried out due to lack of funds (Stornante 2013), but accompanied by orders to keep the shacks under supervision. Even today, in a note dated August 23, 2018, the IACP stated that 96 houses built using funds from the Region have still not been allocated because the houses have not been certified as habitable. The slum clearance was also fraught with difficulties, which included lack of funds, differences between the Municipality and the IACP over who should clear the shacks abandoned by families who had received alternative housing, and about who was responsible for the disposal of the detritus and materials resulting from the clearances, which all translated into absolute immobility on both sides, so that the abandoned shacks were re-occupied by new inhabitants (Tempostretto 2008; Tumino 2015; D'Amico 2017). A further problem concerned the areas in which the shacks had already been demolished, but where the clearance had not been completed and the detritus not disposed of due to lack of funds, thus leaving areas of ruined shacks that were subsequently reconstructed and highly toxic debris (Serra 2017).

As was highlighted in detail throughout Chap. 4, the rhetoric surrounding "rehabilitation" has often been used instrumentally. It has been the means for moving the "needy" population from one area to another, producing the paradoxical effect of creating shanties in new areas, consequently labeling these as further areas to "be restored" (with all the speculative interest that implies). Otherwise, the rhetoric has been used to justify new building expansion made necessary by a fictitious chronic housing shortage, as evidenced by the persistence of the shacks. Finally, it has become a banner to be unfurled at election time in order to obtain the consensus of the people involved. But behind the rehabilitation rhetoric, the city is undergoing a continuous process of deterioration of the urban fabric that is accomplished through spatial and social hierarchization and by pushing the poorer quarters, the "shacks", the slums, and their inhabitants, to certain areas of the city and into certain types of housing.

Thus, applied policies were endless: from the early Fascist reconstruction to the present, with entire neighborhoods placed in the city center still waiting to be "cleaned up", and others that, albeit renewed, have led to the forced displacement of the population in the most peripheral and deprived areas. Quarters in which the lack of services and green spaces are associated with the "separation" from the rest of the city; a city that creates and fixes the limits of a lower-income and disenfranchised population with no other option available.

Due to the opportunistic interests of a specific social class, public space appears as a "denied" space that entangles and implies asymmetrical social relations: relations based on patronage, which transform rights into "favors" and citizens into "clients". If this process is the outcome of public policies the origin of which is very clear, public discourse translates it in the terms of a "shanty culture" (Pruiti 2010; D'Amico 2017) based on the notions of "immobility" and "inability" to be, or to act, any different (so that, according to such discourse, even the external façades of public houses are "re-written" by the dwellers and, thus, degraded). Subjected to this discourse, and to the symbolic and spatial consequences that it implies, the individuals of the margins put into being social mobility strategies aimed at owning their own houses and to provide presentations of their Self that match their own interpretation of a middle-class lifestyle and appearance.

Segregation, however, appears as the most realistic horizon for action. From one dilapidated area to another, from shanty-towns to social housing in decayed neighborhoods, or to illegal self-building in neglected areas. That is, in areas where the State is absent most of the time, but in which it suddenly appears to expose its power and affirm "legality" by means of judicial seizures. An endeavor that, invariably, causes the activation of new cycles of spatial occupation and the continuation of life as it has always been.

By means of their obstinacy to replicate such strategies and to reoccupy public space, they transform marginality into re-appropriation. Urban space, thus, is newly configured and a different rationale replaces that of formal institutions. Nevertheless, the residents' efforts can be normalized and a frame of institutional indifference toward their living environment delimits these spaces. If on the one hand there is abandonment, on the other, there is a "heterotopia" of urban space, resulting in the irregularity of forms, colors, and windows. A characterization of space and an aesthetic affirmation that challenge regulations and imply a potential subversion of the status quo. If such practices are commonly stigmatized and read as an expression of incivility, in our view, these endeavors show the substantial integration of the population of the margins in the ideology of their time. In other words, these are attempts to pursue social mobility, well-being, and even beauty by available means.

6.2 The Shack and the Neighborhood: *Talking* on Spatial Segregation

As described in Chap. 4, the marginal areas of the city developed alongside the main torrents, from the Peloritan hills which lie behind the entire city, or are situated in the old riverbeds of the ancient city streams running down from the mountains, or *torrenti*, in the city (Giostra, Camaro/viale Europa, San Filippo, Gazzi) which had already been covered with cement during the 1950s to make way for the large perpendicular access roads which run from the mountains to the sea,[2] rising up to the surrounding hillsides above Messina (to the so-called villages).

Figures 6.1 and 6.2 show some examples of such combination of shanties and projects in the peripheral neighborhoods.

Figure 6.3 is an example of a local shanty. These structures are one-story rectangular blocks, with insulated roofing panels, which rest on the ground and are exposed to moisture. For this reason, shanties need frequent maintenance of both internal and external walls, and they are often re-painted (as in the case of the white shanty portrayed in Fig. 6.1).

Fig. 6.1 Shanties of "Fondo Fucile" seen from above, with public housing blocks in the background. Photo by D. Farinella, 2019

Fig. 6.2 Old shanties in Viale Giostra. Public housing blocks in the background. Photo by D. Farinella, 2019

Fig. 6.3 Typical shanty, Viale Giostra. Photo by D. Farinella, 2019

As anticipated in Chap. 4, since the 1970s, many projects have been built in the southern area of the city. Today, this section has become a mixture of legal and illegal housing. Self-constructed homes, projects, and large residential complexes with no amenities and services accentuate the impression that this is a "dormitory area" (see Chap. 4).

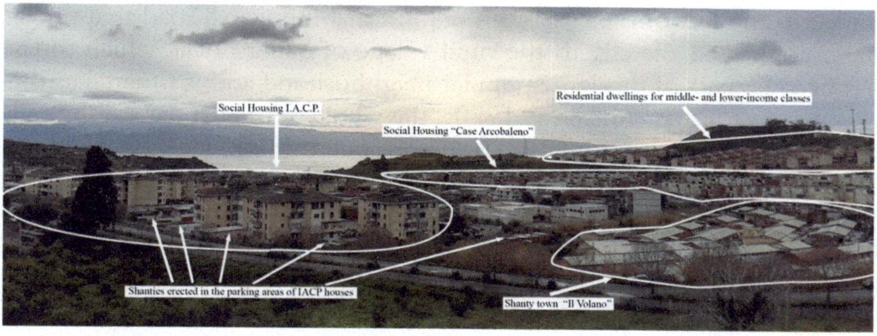

Fig. 6.4 Social housing and shanties in the Santa Lucia area—a view from the speedway. Photo by D. Farinella, 2019

The area of Santa Lucia sopra Contesse is a good example of this peculiar process of social segregation (see Fig. 6.4).

The Santa Lucia area lies "hidden" in a valley that separates it from the rest of the city. Over the course of time, urban intervention has accentuated this symbolic spatial separation. On the left-hand side of the valley, the hills (where there has been uncontrolled urban development) define the area's natural boundaries. Above the area, the shadow of two huge motorway flyovers effectively block any continuity to the hills behind (generating a visual fracture between the urban and rural space). A motorway exit descends directly to the sea, with two wide carriageways spanning the *Torrente* San Filippo, itself already covered with asphalt. These two concrete roads mark the material boundaries on the right-hand side, defining the perimeter of an area that is physically "hidden" from the rest of the city and also defining its symbolic separation. Within the imaginary of a rational, functional city, organized on the basis of the productive needs of the middle classes, this is a ghost area, a mass extraneous to the urban core and one which has grown abnormally and with no apparent rational function: it is not an area destined for commercial activities or services and is not used as a transition or connection zone to other parts of the city. The two main attractions in this area, the football stadium (which is also a concert venue) and the external municipal offices (located inside the stadium buildings) are situated in another part of the valley, separated from the rest of the area by a motorway junction. They are external to the material and symbolic perimeter of the area, and to reach them, it is necessary to take a motorway link which does not actually connect to Santa Lucia, which is linked to the rest of the city by an old internal municipal road that previously connected the city to the rural village of Santa Lucia, which today has become part of this large popular neighborhood and is situated at the entrance to the area. Since the 1970s, the whole of the flat part of the valley, which was once grazing land or used for cultivation, has been overrun by large-scale social housing projects: forty public housing blocks were initially built by the competent public authority, IACP (around 350 apartments) (the yellow blocks in Fig. 6.4); successively, at the end of the 1980s, the Arcobaleno houses were added, many of them allocated to the residents from the shanties in the *Torrente* San Filippo area that were cleared to make way for the construction of the new stadium; and lastly, in 2011, a large new complex of forty-four houses was built.

As shown in Fig. 6.4, in addition to the social housing, there are also illegally built houses, shanties, and self-builds, situated along the roads on areas of abandoned land, even in the car parks attached to social housing, as well as inside a small shantytown called "Il Volano". These buildings are tolerated and even encouraged by the residents, who are thus able to resolve their children's and family's accommodation problems. This underlines the existence of close relationships based on reciprocity, which help to fill the gaps left by the shortcomings of the welfare state and public action.

6.3 At the Border Between Formal and Informal: Strategies to Get a Home

In dilapidated neighborhoods like Santa Lucia, it is very difficult to document the diversity of the residents. This population is often made up of low-income families, much larger than the average. Members of these households have extremely low educational qualifications. The head is generally unemployed or employed in low-skilled manual jobs. Frequently, he is on the dole or receives a type of social pension. The entire household, then, is dependent on public welfare. Under these circumstances, petty crimes are often a necessary evil. Imprisonment and house arrests, therefore, are quite common. Moreover, early family formation and parenthood are also frequent. Lack of work, or prospects, does not interfere with these choices, and many young couples live with one of the two families of origin. Intergenerational solidarity is generally high. For parents, in fact, it is common to help new families, in both financial and relational terms—that is, with regard to expenses, accommodation, and childcare. In the presence of extended forms of irregularity, such forms of support are essential. Public welfare is insufficient: on the one hand, it is a support structure that can be evoked, especially with the presence of children (under the form of maternity allowance or shopping vouchers); on the other hand, welfare is that apparatus that prevents social mobility and causes spatial segregation. Indeed, social and spatial segregations are the consequence of the convergence of local tactics aimed at promoting speculative interests and general social policies that are not apt to support new and old households. Unlike common wisdom, given this framework and the forces that operate, our view is that slum-dwellers do not practice a "culture of poverty", nor are they disenfranchised. Rather, they produce agentive strategies that constitute as many forms of "resistance", apt to claim their existence, against a public and private structure that pushes them to the margins.

Moreover, unlike tactics (see Chap. 5), here, our use of the concept of strategy implies the ideas of "project" and a longer temporal horizon for an action's course. The aspiration, somehow, to stabilization and social mobility to be pursued by means of rational actions. We are cautious in our use of these latter notions as we do not want to write another "romance of resistance" (Abu-Lughod 1990). Likewise, we are aware that conflicts that take place in fields like the one being studied are not really aimed at upsetting existing relations, but at simply keeping things the way they are (Ortner 1995, p. 175). Our use of the term resistance does not imply the idea of "great rejection", nor the belief that one is witnessing examples of "counter-conduct" (Foucault 2007, p. 196). Rather, we refer to a specific analysis of power relations (Foucault 1978), for which informality is a "weapon of the weak" (Scott 1985). By means of informality, individuals and households act in a way that increases their chances of controlling social space and pursuing autonomy in spite of the several obstacles they face over the course of their lives. The pursuit of regular aspirations against one's odds is, therefore, what we call "resistance".

After all, "house" ownership is not only a way to supplement pensions, or of strengthening one's future (Castles and Ferrera 1996), but it is a way to mitigate instability—a dynamic seen in action in many other places (Desmond 2017; Hyra 2017). In the Italian case, moreover, it is also the symbolic manifestation of participation in (and belonging to) a middle-class lifestyle.

In consideration of the structural obstacles that hinder the acquisition of minimal forms of security, we claim that the informal strategies utilized by slum-dwellers to pursue their aspiration to housing are the only practices that allow them to build a possible future. At least in their eyes, the individualistic appropriation of public resources—for example, the occupation of public houses or the alternative use of public space—is the only practice apt to secure them a future. With Lefebvre (1991, pp. 164 passim), we can call this "appropriation", that is, one of the ways through which to exert one's "right to the city" and fulfill both basic and expressive needs.

The building of a small house or shack from scratch, or the occupation of an existing one, is one of the strategies we refer to. This course of action is undertaken consciously—often, in order to enter the urban renewal processes. The occupation or the building of a shack, especially if in areas to be renewed, makes the occupants eligible for the waiting list for municipal housing. Although this trick can take decades, it has created widespread mechanisms of imitation.

> A.—I live in social housing. It is still social housing today. [...]
> I.—So how come you got a house here? I mean, was it assigned to you or did you take over from someone?
> A.—Yes, 'cos we were living in these houses, in these little houses [*the "casette ultrapopolari"*] [...]. It was in the area where my husband lived and they [*the previous assignees*] were selling these houses ... I did it all up, I fixed everything, I even had a terrace then! Later, they came round (the Council) and said it was an area that was going to be cleaned up and they cleaned up the whole area, and so, for me, they gave me this house [...]. (Woman, aged 56, divorced, 1 child, interview n. 1)

Especially if one has lived in a shack before, the idea of returning to a similar accommodation is not absurd. Above all, if such a step is conceived as the price to be paid to obtain a real house—regardless of the risk of being trapped in the mechanism for a very long time. These aspects emerge clearly in the following interview:

> A.—We (the family of origin) were living in a shack [...] I was twenty years old when they gave us a house [...] Then, I got married and I had to go and live in a shack again.
> Interviewer—And why did you go and live in a shack?
> A.—Why? Because my husband has always been a building laborer and so he isn't always working [...] He didn't have a regular wage, like an office worker. That's why we went to live in a shack [...] that we were told was just temporary, but we lived there for twenty-eight years. [....]
> I.—And how did you get the shack? Was it yours?

> A.—We just occupied one. […] There were these shacks […] and we just took one illegally. Then they didn't throw us out and so we stayed for twenty-eight years.
> I.—So you knew these shacks were empty?
> A.—Oh, yes, they were empty. They'd given the occupants housing in Santa Lucia. Then, they left these houses empty and so we went in during the night and we just stayed there. I wish we hadn't done it. […] I regret it. Because it's a life of sacrifices. For what? It isn't worth it. My kids went to school, they grew up […] They got married but I didn't have the pleasure of seeing my daughters leave (dressed as brides) from my house […] In the end, what can you do? […] Now that I'm older and my daughters are married, they've given me this house. Thank God for that at least. But, all the best years of my life have been spent here, making sacrifices. (Woman, aged 58, married, 2 children, interview n. 19)

An emblematic case is the one concerning the *Torrente* San Filippo in Santa Lucia Area. This area was not included in any rehabilitation plan. The decision to build the new municipal stadium there and, subsequently, to expropriate the land (occupied by some illegal houses and a few stables) and compensate the owners with public houses has generated paradoxical as well as predictable effects.

As soon as the rumor began to spread that this might be a possibility, the individuals began to mobilize, filling the *torrente*, the stream, with shanties and illegal constructions.

> Interviewer—Where did your family live before?
> Man—[After we got married] We had this little house. What people call a shanty. […] We made ourselves a shanty to live in. […] You know, in the *Torrente* S. Filippo where, even now, they gave us houses.[…]
> I.—Were you aware of the city plans? Someone came out with this idea?
> M.—This was the idea… seeing that we could see other people building these shanties… like, it was an empty *torrente* and anyone could go there and get a space to build a house, and so everybody was doing the same thing. This was the main idea anyway. Like in two years, this damned *torrente* got all filled up with shanties […]
> I.—But people knew that the city would have given the house sooner or later?
> No, no, no! No one knew […] Maybe we was a bit lucky, 'cos they didn't give us these houses 'cos they had to, but 'cos they wanted the space to build the sports center. […] It was an amnesty. And they did it. (Man, aged 49, married, 3 children, interview n. 14)

Another interviewee recalls:

> My husband had a sort of "garage" then (in the area of the *torrente* which was expropriated), and they were evicting everybody at the time, they gave us a house, a small prefabricated house [...] It was just temporary, [...] then [...] they assigned us a house. (Woman, aged 54, married, 3 children, interview n. 6)

Another common method to obtain a "house" includes the unlawful taking over of IACP housing. By means of the illegal trading of contracts and the shifting of residence to the house of an assignee, public housing lists can be effectively bypassed and one can qualify for a house.

> Interviewer.—How did you get this house from the Public Housing Institute? Were you living in a shack?
> Husband—No, no. I had bought a contract, at that time. [...] From a woman who emigrated, and then this poor woman died too! We won't go into that! [...] Anyway, there were a few situations with the house.
> Wife—We had a few problems!
> Husband—Because, you know, the house wasn't mine, it wasn't allowed. So I called them up over there [...] (the daughter)
> Wife—Because buying a contract isn't allowed.
> Husband—And she says "my mother is dead!" So, I spoke to someone, "we can deal with this", he said, "get them to send a death certificate and we'll try to sort it out ...". But do you know what I had to go through?! [...]
> I.—So, who was it who actually helped you with the house?
> Husband—First of all, we got a lawyer [...]
> Wife—And then a clerk from the Public Housing Institute.
> Husband—And then someone I met through a mate of mine [...]
> Wife—And he helped us out.
> Husband—And they sorted out the situation for me! (Man, aged 55, married, 2 children, interview n. 13)

Where there is available private or municipal land, another strategy would be to build an unauthorized house, usually next to the parents' house.

> A.—[...] we built our house very slowly, making enormous sacrifices! My father gave me a piece of land next to his and we built our house there. [....] My mother helped us out by cooking for us. I would eat there, at my moth-

> er's house, and I slept at my mother's too. [...] We managed to build this house with what my husband earned. After we'd finished building, we then had to furnish it.
> I.—And how long did it take you to build it?
> A.—Three years, three years. Then we had to furnish it. I furnished it by paying in installments. Through this guy from Catania. In those days there were these people from Catania who went around the houses and you paid an amount, say twenty, thirty a month, whatever you had. [...] It's a nice piece of land where we built, with my brother and my other brother. I live downstairs, my brother upstairs, so all three of us put in the money together. (Woman, aged 47, separated, 1 child, self-build house, interview n. 20)

Indeed, some of the previous tricks have been impeded by the new urban policies. Later changes in the legislation and the city regulations made requirements more stringent. In particular, the Regional Law n. 4/2002, which provided instructions on the rehabilitation of the city of Messina (and that replaced the previous Law n. 10/1990), established the right to public housing only for the households whose members had resided in the areas subject to rehabilitation for at least three years, during the years December 31, 1995–December 31, 1998. Furthermore, these households must have resided in the shacks, and provide evidence of this, in the moment that new apartments were given (and not only assigned). Those who took possession of a shack after the considered period were excluded from these measures. The only exceptions concerned the newborns who came to light at a later date from a residing family, and the new couples whose members included one subject who had resided in the unit during the period 1995–1998, and who lived in the shack together with the recipient household and his/her newly formed family. At least partially, this latter norm explains why households surveyed in the rehabilitation areas are, on average, sensibly larger than in any other part of the city. Moreover, the three-year rule made the trading of shacks less convenient. Furthermore, this rule drew a line between those who dwell in areas to be rehabilitated, those who have the right to public housing, and those who are excluded from the lists. This being a line, or a confine, which is problematic and ambiguous, as will be shown in Chap. 7. In addition, those who occupied public houses belonging to IACP were excluded from the waiting lists for five years.

However, in spite of these changes, the "struggle" of these households to secure themselves a better future continues. Space, in other words, is still used in creative ways and beyond the norm. With regard to newly constituted households, the role of the family of origin and family networks are central to the activation of non-material and "free" resources. Family members often make economic contributions to the purchase of the house, or make other resources available (grandparents' houses, building land), help with the reconstruction or with the building of brand new houses—which is often done with the expert help and knowledge of family members who work in the building sector, according to the model of the mutual exchange of labor (a "non-market" economy which was also widespread, e.g., in some rural, socialist Hungarian villages when it came to building one's own house. See: Sik 1988).

> It was a tiny house, my own house, that my mother gave me, social housing, so I didn't have to pay for a house! [...] well.. at least I had that, a roof over my head!. (Woman, 60 aged, married, 2 children, interview n. 21)

> A.—I got married in 1991 [...] My partner was pregnant. I told my dad: "even though I don't have a job, I want to get married!" [...] My dad said: "don't worry about the baby girl". The milk, the diapers... he bought everything.
> Interviewer—So, she was pregnant when the two of you got married...
> A.—Yes, yes. We had an escapade. We left our homes for a few days, you know...Of course, I asked myself "how do I live now with a daughter?" My dad said: "don't worry! If two of us can eat, five of us will eat as well!" And in fact we ate at my dad's place, and he also gave me some money to go out and have some fun! [...] He said: "In case, you'll eat at my place", both me and my brother.
> I.—Was your brother in the same situation?
> A.—Yes, the same thing. He was married, and had one child [...]
> I.—Where you and your wife lived at that time?
> A.—I am telling you. My brother lived in a house of the mother of my father, and me at the house of my other granny, the mother of my mother. (Male, aged 44, married, 3 children, interview n. 16)

The family therefore seems to act as a safety net that helps to generate intergenerational economies of reciprocity, whereby children and parents take care of each other.

A.—I bought the house in '88. Yes, because I was living with my mother.
I.—Was it social housing?
A.—Yes, an old building. I lived with my mum, then my sister's husband died and she went to live with my mum too. Well, you know, the house was pretty big, but ten people, it wasn't great living like that. [...] Anyway, to settle things once and for all, my husband says "let's get a house!" But then we had some problems because my husband got depressed. Well, you know, you have to buy stuff, furniture, and this and that, and we had a few problems and you know, sometimes, I had a few too.
I.—And was it social housing?
A.—Yes. It was social housing, a bit old and the owner was selling it 'cos she and her daughter moved to another town. It's not like they were new apartments! (Woman, aged 52, married, 1 child, interview n. 1)

The outcome of the strategies aimed at solving the housing problems consists in the generation of mobility and spatial flows within the territory of the margins. Such flows, and the whole of the tactics that compose them, stand for forms of activation, agency, and adaptability that are very distant from the portrait of the passive, idle, and welfare-dependent southerner which is at the center of the modern national discourse.

As discussed earlier, life in shanties is the result of structural forces linked to the economy of space and to insufficient welfare. If we lay aside dependency culture rhetoric, we can observe how these housing strategies are the result of imitation, efforts, sacrifices, and reasoned choices. The subjects plan their lives and mobility around the construction of their own autonomous family. Within this framework, the collected interviews highlight that the problem of accommodation and housing presents itself in all phases of family life (marriage, separation, death) and show that the boundaries between social housing, self-builds, and shanties are fluid. Just as the boundaries between "legal" practices and informal and unconventional strategies to gain access to housing are fluid. The choice between one and the other depends exclusively on practicalities, on the analysis of opportunities, on the availability of money to pay a deposit or rent, the possibility of finding some abandoned land on which to self-build a house and the freely given help to build it, and so on. Individuals and households who sketch unconventional strategies are the same people who have tried regular channels but were not allowed to do so. Their actions stem from either objective obstacles or from their concluding that it would be impossible to comply with legal avenues to housing. It is then the feasibility of the solution that pushes them to opt for a formal or informal course of action, or a mixture of both. Due to the nature of their unstable jobs, usually, they do not qualify to get a loan. But even if they do and obtain small sums, a shanty or a dilapidated house is the only property they can aspire to. This was the case also for one of our witnesses:

> A.—No, I didn't buy the house. [...]. I was living in Via Palermo, renting a house. After, my parents' [...] my mother's mother (a visually impaired grandmother) and my invalid uncle who has no legs, needed someone to look after them and so they convinced me to move in there. And I definitely went through a lot! [...], I didn't have a life. I wasn't free any more. [...] I felt conditioned, forced, because I had to do so many things, shopping, cooking. That was my life [...] Then, in fact, out of necessity I built a house with a loan, and I was living with my husband then and he paid it, with a loan I built two rooms illegally, after that they gave me a house, which is now a council house which I pay monthly.
> I.—What do you mean? You built two rooms next to the house you are living in now?
> A.—No! No! In an area where everyone was building! It was illegal land. [...] I was so desperate because I couldn't carry on living like that, so we said 'you know, they're building houses ...' and we built it with help, we didn't pay any workmen though. Just the materials that we bought with the loan and they built me this house. And afterwards I felt much better!
> I.—And then you were part of the redevelopment [...] How is it that you were given a council house?
> A.—Because they had to clear away those houses because they were going to build the sports ground. [...] The 'San Filippo'. I lived there, in that area. And, from morning till night, I remember the year I was living there [...] it was in 1990, when I was working in neurosurgery, I got the phone call, they said 'the local police are here, they're demolishing the shacks, they're giving us houses!' (Woman, aged 53, separated, 2 children, interview n. 21)

It is problematic to ask to what extent the choice to "occupy" illegally is the expression of a real need, a lack of alternatives or rather an opportunity (either real or imagined). On the contrary, even in this latter case, in the daily fluidity, the one is the same as the other. As we noticed, the families who decided to unlawfully take over public housing or to occupy a shack had also tried more "traditional" routes such as renting, and continued to do so, for example, asking for a loan to "renovate" the shack, which represented their only chance of an investment.

Moreover, in the act of choosing which strategy to pursue, actors face various dilemmas. However, the careful consideration of means and ends shows the substantial adhesion of this population to the utilitarian logics typical of capitalist ethics. Individuals are aware of the costs as well as of their liminality. This latter being the condition that puts them at risk and makes them vulnerable to the gambling of other stakeholders. That is, to the potential costs of risky transactions that take place in informal and unregulated settings, exposed to frauds, and free riding.

At the same time, the actors are also aware that that liminal context provides them with the same chances to bend the rules and, somehow, to level the playing field. However, the frame remains uncertain, ambiguous, and unpredictable: a positive or negative outcome, far from being adequately provided for (albeit rationally planned), was often the result of unexpected situations, as well as the individuals' abilities to act. Unexpected events were always just around the corner, ready to materialize, such as the sudden death of the house owner for which one risked losing the right to take over the house contract (as the man relates in interview n. 13).

For this reason, the ability to mobilize relational resources and forms of reciprocity based on trust are important. The actors know that some properties (including a contract with IACP or one for the acquisition of a shanty, which is not habitable and cannot be legally sold) are not for sale, but that only the right to live in the property has been transferred.

Within this context, the counterpart, that is, the one who sells something that cannot be bought, constitutes the first problematic aspect of the transaction. Will the pact be honored? Will it be possible to rely on the brokers? These are the questions that the actors pose each time to themselves. The relational context of the witnesses, thus, is populated with "ghosts" and ambivalent characters. There are, for example, the "spies", that is, individuals who have interests at stake and follow, or affect, the outcomes of a transaction. And there are the "friends" who propose themselves as intermediaries but who can be either insidious or honest.

> A.—[When we got married] we were renting this place in Camaro. And then they gave us this house from the independent institute in Bordonaro!

> I.—ah, and how come they gave you a house?
> A.—'Cos we also had an old house and now and then we went there… and then, there, they was giving out houses.
> I.—Right, so you were renting the house where you were actually living, but at the same time you owned a house … a shanty ….
> A.—….Very very old! We had our residence there. You could do that then, and so they gave us the house, but thirty years ago! Eh, it's not like it is now … thirty years ago!
> I.—But this house that qualified you for social housing, who did it belong to originally?
> A.—Then they were selling (the house) and so we bought one and we'd always rented, because this house was old and humid. Then someone ratted on me and I had to go and live over there for a bit. Then, when the employee for the Census passed, he certified that I was a resident there, so they included me in the recovery plan and I was entitled to the public housing. But, you know, thirty years ago! thirty years ago I was young, two kids, we had nothing else… it was mine, that house! […] But it was so damp, really damp… so my husband paid rent in another house!
> I.—Ah! But lots of people did that?
> A.—Back then yes! Then all you had to do was buy a house, a shanty … and they gave you social housing! Of course, you couldn't have any other houses…. You weren't supposed to have anything and we certainly had nothing! (Woman, aged 60, married, 2 children, interview n. 23)

A second aspect is setting in motion pragmatic and calculating behavior in order to answer the question: what and how much is the individual willing to risk in order to realize their dream of owning a house? In a context in which there is a low level of regulation, where all are potentially "illegal", it is worth the risk. This is amply shown in the interview below, where the interviewee informally bought a housing contract so as to live in the "Case Arcobaleno" social housing complex, with their family of origin.

> A.—I've been living here for twenty years! My father was assigned a place in the Case Arcobaleno. […]
> I.—Why was he assigned a house?
> A.—What do you mean, why? Because there in the torrent, they got rid of everybody so they could build the stadium! My dad was living in a shanty where there's a football pitch now!

I.—Did he build the shanty himself?
A.—Of course! But it wasn't a shanty, it was a house! These are shanties (the Case Arcobaleno, n.m.)!! They called them shanties! But they weren't!
I.—But wait a moment—were they authorized to live there?
A.—Ah?!? But no one, no one was authorized! And how did it end up! They put them all in social housing! Now the same thing is happening in XXXX, where there are houses that are all illegal, and they have to demolish all of them and they're giving them social housing!
I.—Ok, but you're illegal too!
A.—But I'm another type of illegal! Actually, now I know that I'm illegal too. I thought I wasn't illegal! Then I realized that I was illegal! Because the owner who gave me the house, he's got his residence there, he's the assignee. So, if I'm with the assignee, I can stay, no problem! But now the assignee's got another house, he bought it (so his right expired) [...] But it's not like I knew anything about it, is it? [...] (I bought the house) through a friend, who told me "You know what, he says, Ciccio's dad's selling his house!", "what d'you mean he's selling his house?"[...]. Well, that was it. I had to have that house![...]
I..—But what on earth is he selling, if it's not even his!
A.. Yeah, but he's letting you have it! You're gonna get it! Maybe it'll go alright for you, and you can stay; maybe it won't go good and you'll have to leave it! But, you know, a bit like me right now, it's going good for me! When it goes wrong, I'll just say "ok, so it went wrong!" [...] To get in, most of them, some wanted 25, some wanted 20, some wanted 30 thousand euro. You know how much I paid? You couldn't even buy a car with what I gave him! [...] 7 thousand euro I gave him! And the thing was, he had to leave his residence there, he had to leave it there with me for at least a year. He had his residence there [...], but before I went there he had to take the keys to the Town Hall, 'cos they saw he had another house like. But I didn't know nothing![...] He knew alright! [...] 'cos the papers—that said he'd lost the right to the house—they'd already arrived. When I went to the lawyer, this one saw all the papers and he said: "no, they sent you the eviction notice three years ago, so it's useless you going to a lawyer". And thank god this lawyer was a friend of my dad's! (Man, aged 30, 2 children, married, interview n. 22)

As claimed by the previous witness, it is the cost of a similar house that makes this a convenient and, above all, possible investment.

6.4 Living at the Margins: Between Public Constraints and Resistance

These strategies demonstrate different ways of achieving access to urban space that is "denied" but that becomes physically "lived-in".

The subjects using these sorts of strategies are perfectly aware that they are not really buying a "house"; they are trying to underline their "right" to a house through the State welfare system, either by living in a shanty in an area designated for rehabilitation (even if they are not in possession of the residence requirements mentioned in Chap. 4) or by sub-letting social housing (paying a sum of money informally to the assignees). These are all operations that can help them to access public space in order to gain from its "benefits". They try to be more "connected" than before, and it is difficult to "throw" them out.

This right to "house" ownership is built on an individualistic basis and is one of appropriation through illicit actions: by appropriating a public space, the families "act out" their presence within the territory and their condition as "disadvantaged people", and in doing this, they make it a "reality", they "actualize" it. As the years pass, remaining in a disadvantaged situation transforms what was initially appropriation into a right, when the "sacrifices" that have been made to earn the right to a house compensate for what is actually an illegal act. These aspects are examined in depth over the course of the next chapter.

For now, it is important to remark that the strategies aimed at acquiring a house are "circular" for being limited to certain neighborhoods and dwellings only. Mobility paths conduct this population toward specific, dilapidated areas of the city. Their practices, strategies, and tactics, moreover, are mostly illegal or semi-legal. For people with irregular incomes, in fact, available opportunities are often at the edge of the law. The illegal renting of social housing and the acquisition of houses located in remote villages where prices are low and the general conditions extremely bad (with mold and humidity, broken fixtures, no heating or elevator, and water tanks made of fiber cement and, thus, toxic) are some of these opportunities.

These aspects make clear that people living in the popular neighborhoods are trapped in a spatial segregation product by public policies. In their relationships with public institutions, these subjects experience the ambivalence and the asymmetries of power: the institutions' bargaining power often ends up bringing the action back within a framework of indifference and noninvolvement. Compared to the individual choices, public action once again confirms its power, relegating people in a shanty or in a social house made of low-quality materials, or in a new popular "ghetto" district, without services. So families are destined to spend the rest of their lives in suspension, waiting or hoping for better housing, in a temporary situation that becomes infinite and permanent, and that produces existential precariousness (Giddens 1990).

> Now we've got social housing, it's an old house, like, you know, public housing [...] it was my gran's [...] eighty years old, old house, really old, but we put up with it [...] old houses they are, you know, public housing, we don't even pay now, we been there for forty years, but they don't give us no houses [...] my gran's dead, my dad's dead, I'm gonna die too... when are we gonna get one? It was my gran's! [...] Then it got passed on to my dad, then my dad died and it got passed to us. (Woman, 57 years old, separated, 2 children, interview n. 23)

> I have to thank God I've got a roof over my head. But, there are a lot of things to sort out, like the floor which keeps coming up [...] If you came to my house, you know, it looks quite nice, welcoming because we try to keep up with the maintenance, in our own little way, even a tile we stick it back on when it comes off. [...] You know, for the materials they didn't spend very much [...]! (Woman, aged 53, separated, 2 children, interview n. 21)

The next excerpt shows how, in response to institutional abandonment (and perhaps in contraposition to it, as suggested by the recurring use of the pronoun "they" to designate the institutions), the residents shape and produce the space, which is readapted and newly regulated in accordance to personal needs and to logics that are locally shared and applied in opposition to the formal one.

> A.—The Arcobaleno houses, the famous Arcobaleno Houses that they gave us in 1989. [...] We've been here for nineteen years now, and we've never paid anything, [...], we've never had a contract. Now they're making some repairs, because, you know, it's impossible living here, there's leaks and so on. I don't know, I don't know if this is good or bad, I don't know.
>
> I..—But they say they have to demolish all the illegal building that people have done...

> A.—That's what they say, but what they don't say is that this is an area that is totally neglected, so you get drug addicts coming into your house at any moment. They say we've done things… […] Well, I tell you what […] just three days after they gave us the houses, I spent 5 million, then, and six-hundred thousand lire on a reinforced door and we put security doors on all the balconies and windows. That's the first money I spent there. The Bronx was nothing in comparison. […] They say that we did it, and I was one of the first, I was, because it was like no-man's-land. This is what happened, 'cos it wasn't run by anyone. And it's like fifteen years that there's no light on the streets, 'cos they say, "well they steal it!" (*the residents illegally connect to electrical power lines through the street lighting*), "yeah, okay", I mean, not because there are people who steal it—because it's true—I don't see why people who haven't done anything have to pay! (Man, 49 years old, married, 3 children, interview n. 14)

Perhaps paradoxically, the resistance to the institutional and enforced marginalization is expressed through illegal building. This practice generates a new interpretation of urban space, and it also responds to that "anguish" that, as De Martino (1973, p. 95) has noticed in a classic book, expresses the willingness to exist as a presence in the face of the risk of non-existence. By means of the self-building of a single house, or the widening of a balcony, the individuals reclaim property as well as space, and they project their presence in the territory. Moreover, even if the middle-class glance tends to see evident signs of degradation and havoc in their own interpretation of space, for the residents, this is simply "home":

> My dad lived in a shanty […] but it weren't a shanty, it was his home! These things here (the Arcobaleno houses) these are shanties! They called'em shanties! But they weren't! (Man, aged 30, 2 children, married, interview n. 22)

Thus, not only shanties can be homes but projects can be shanties. In a sort of *continuum*, the individuals' practices of resistance are never blatant actions of struggle. Rather, they consist of attributing meanings that are subversive with regard to the *status quo*.

Like the Iranian "underclasses" investigated by Bayat (1997, p. 7), engaged with their "quiet encroachment of the ordinary", usually, people of the shanties do not employ organized and collective forms of mobilization to claim their right to exist, nor affirm alternative values (even though, occasionally, they do).[3] It is also possible to argue that, in the case of these actors, their practices outmarch ideology, even though this does not imply that they do not have political visions, or that such visions cannot compete with the hegemonic ones. However, at a superficial level, one notices certain paradoxes that only an emic knowledge of the meanings conferred by the subalterns to their actions can explain. In other words, to develop an analysis, it is necessary to leave out normative approaches to the subject matter and deploy, rather, perspectives entirely centered on the actors, their processes of signification, and their intimate ways of assessing the surrounding world.

In the light of these considerations, our idea is that they use the term "house" to describe their shanties, in open conflict with the categorization of space created by public action—similarly to what Barrat and Green (2017), among many others, have observed in similar contexts. Homes are made habitable and comfortable. Maintenance is carried out regularly (Figs. 6.5 and 6.6), often due to the fact that at least one member of each family is "in the trade". Generally, the inside of the houses show much care for detail, with chosen tiles, wall paint, and furniture, inspired by current fashions (Fig. 6.3). Moreover, shanties and projects are, obviously, gendered spaces (Spain 1993), which, nevertheless, fulfill expressive needs for both sexes (Thompson 1994). A "space of expression" that, among other things, substantiates the idea of appropriation as a manifestation of relatedness, urban culture, and living memory (Jiménez-Domìnguez

Fig. 6.5 Old shanty "refashioned" by the inhabitants; on the right, shanties with the courtyards fenced by the public sidewalk. Photo by D. Farinella, 2019

Fig. 6.6 Old "Casette" in Camaro (Torrente Zaera) restored by the inhabitants. Tangled up electric wires suggest the presence of illegal connections to electricity supply. Photo by D. Farinella, 2019

2007, p. 99). And that, not by chance, provides the observer with the opportunity of reading even these spaces as a text—similar to what happens to the most integrated or highbrow neighborhoods of the cities (Stevenson 2003, p. 54 passim; Zampieri 2018).

Though this type of space has not been chosen, but imposed decades earlier, space at the margins is somehow "loose" (Frank and Stevens 2007). If the notion of "looseness" applied to urbanism implies the possibility for public space to be altered for unintended uses, the considered neighborhoods of Messina are, similarly, spaces within which it is possible to exercise creativity. Public spaces are constantly invaded, re-appropriated for alternative uses and, certainly, even privatized. Former nursery schools that have never been utilized can be transformed into houses. Courtyards can become winter gardens, vegetable gardens, garage, and laboratories within which different activities are carried out (Figs. 6.7 and 6.8).

Fig. 6.7 Social housing "Case Arcobaleno" in Santa Lucia "refashioned" by the inhabitants. Photo by D. Farinella, 2016

Fig. 6.8 Social housing "Case Arcobaleno" in Santa Lucia "refashioned" by the inhabitants. Photo by D. Farinella, 2016

"Shanty culture", rather than the "culture of poverty" (Lewis 1966, p. 21), is the formula that local public discourse utilizes in order to describe the ideology that accompanies and follows urban decay. But the words of the residents of these dilapidated areas tell a different story. In relation to the practice of giving sections of the city almost for free to speculators who have built on every inch of land, or used the land for open garbage dumps, an illegal self-build is an example of enhancement rather than the contrary. Self-builds, in this perspective, are a way to achieve one's right to home ownership and solve, as said before, that "anguish" (of non-existing) that accompanies the lives of this population. In the clash that opposes them to the City, and in the institutional refusal to acknowledge that shanties are homes, this very same anguish is continuously re-activated, together with those tactics and strategies that, on the one hand, shape the space and, on the other hand, "trap" the dwellers of the margins. The heterotopias that in our view characterize these spaces—and that appear composed of conservatories located in municipal streets, in ovens built on verandas, and the illegal courtyards where residents make *focaccia*, in the colorful roofing and in the other "extensions" to illegal property—are therefore fed both with anguish and desire (see Chap. 7).

But this is also the erection of a "city of thresholds", which "symbolize and concretize the socially meaningful act of connecting while separating and separating while connecting" (Stavrides 2007, p. 176). An alternate (dis-)connection, as we have seen over the course of the book, which has been actively pursued by the constitutive logic that led the new city formation process, and that entails the "dependent" social and economic relations described in the previous chapters. Yet, in Foucault's words, heterotopias are places inhabited in ways that deviate from what society considers and imposes as normal and that presuppose a system of opening and closing (Foucault 1993, p. 422; Stavrides 2007, p. 177). And while, unlike Stavrides (2007, p. 178), we do not believe that heterotopias necessarily diffuse a "virus of a change", we argue that the neighborhoods of shanties and projects we studied express, nonetheless, peculiar practices of "resistance" and demands to be acknowledged within that political and public arena where they are, in fact, unrecognized. Far from any form of romanticism, these spaces are practically re-appropriated and reshaped in a new informal regulation between the individuals, where, still, an individualistic frame prevails.

Following Scott (1990, p. 108 passim), moreover, it is possible to claim that informal spaces such as the one here investigated, tend to become the expression of an involuntary "dissident culture": a culture that is not actually so different from the dominant one, but different enough to not resemble it, and also to generate in those who belong to it a range of negative feelings that are largely caused by the awareness of their position and options in society. A "resistance", however, that is not intended as such by the actors yet it is constructed and recognized as threatening by the dominant counterparts (Hollander and Einwohner 2004, p. 545). And it is here, in this fracture between social worlds, that the informal space becomes a space for subaltern "politics", expressed through languages, styles, ideas, passions, and, above all, practices, that are either partially or radically different from the dominant ones, and that appear as the only viable solution to the problem of making ends meet and guaranteeing the right to life and joy (see Chap. 4 on this latter point). The exact opposite, then, of what the local civility discourse emerged after the earthquake claims.

6.5 Conclusion

In a book on some Italian recent disasters and the speculative processes they generated, the author describes the outcomes of the CASE (antiseismic sustainable and eco-friendly housing) Plan implemented in the city of L'Aquila in the aftermath of the earthquake that hit the area in 2009 (Caporale 2010).[4] For the Plan, over 22,500 people had to be permanently hosted in wooden houses or in pre-fabricated multi-story buildings within eighty days of the seismic event. In short, this was the cottage model, based on barracks, that had been applied in Messina in 1908.

A model that has, not insignificantly, remained part of the urban landscape over a century later. Although the barracks are not exactly the same as they were in the past, those first housing units generated forms of social organization, control, and exchange whose negative features were heightened by the insufficiency of local and national welfare, and which became enduring characteristics of the place. In this sense, the organization in post-earthquake Messina anticipates at least one of the great contemporary Italian disasters, also in reference to housing solutions and the future prospects of the individuals, families, and places involved.

Barracks were integral part of those territorial segmentation policies that have been implemented and pursued by different political regimes—from the early Unitarian State to the Fascist regime and the Democratic Republic—and form the basis for current urban stratification. At the basis of this process, there was the so-called shanty relocation, which consisted in transferring all the barracks and temporary housing present in central areas to more peripheral city sections that were excluded from the emergency plans and the then new master plan (Piano Borzì). As we have seen, this is how today's peripheral areas, in a geographical and/or social sense, came into being.

Reconstruction and the building industry as a permanently leading economic sector became, in the city of the Strait as well as in other Italian urban centers,[5] the outlet for the needy masses resulting from wars, disasters, internal migrations, and speculative processes (Ferrarotti 1970). Reconstruction became also the tool through which exercising political control on such masses, which were ever more dependent on particularistic relationships with the local political, economic, and religious elites. With regard to disasters, these observations suggest that these events are characterized by *longue durée*. Not by chance, "catastrophe" is the word we utilized to highlight the *temporal* element implied in certain disastrous events. In the analysis of disaster-related time lines, however, we argue that it is important to include the role of the political structures that precede, accompany, and follow undesired events. For example, in the case of Messina, the post-unitary liberal setting within which the earthquake took place is not alien to the outcomes of reconstruction. The city had been destroyed before, but it was caught in the cycle of marginalization only during this last crisis. The new unitary regime had multiplied the number of actors and the levels implied in the exercise of power and, therefore, in the allocation of resources. Inefficiencies and instrumental uses of resources were now intertwined. Particularistic uses of power, the creation of clienteles, and the concentration of wealth in the hands of specific classes were more important than an effective reconstruction.

In addition, continuity in marginalization marks the recent history of the city, including the public sector (transport, garbage collection, public finance etc.). Nevertheless, perhaps contrary to what common sense would suggest, over the course of the decades, massive expansion and territorial erosion took to characterize this chronically depressed city. This has added a "physical" issue to an enduring social question, in terms of floods, landslides, and exposition to "natural" catastrophes, as shown by a mudslide that, in 2009, killed 34 people and left homeless hundreds of people in peripheral sections of the city (Giampilieri and Scaletta) (Falconieri 2017).

Therefore, in opposition to popular discourses, our goal was to show that shanty-towns and social housing areas are not peculiar and disorganized moral spaces that follow a logic of their own. Rather, in our view, they are a *produced* space (Lefebvre 1991, pp. 82–83). Historically, patronage has been the channel for this subaltern form of inclusion, and housing and urbanism are the main tools utilized to build bonds that linked lower classes to the local centers of power. Urban planning, exemptions, derogations, and other similar "tricks" were the instruments utilized to control different groups and classes, which included landowners, developers, poor people aspiring to public housing, and also precarious and unskilled workers whose inclusion in the labor force passed through the construction industry. At an ideological level, "hope" is, in Chubb's words (1982, p. 167), an important element in sustaining a clientelistic system of power in a context of limited resource. Hope, moreover, is related to temporality. It implies the notion of waiting and a sentiment of faith in someone. Not by chance, in opposition to the model of the "rational voter", for decades, people continued to vote for a politician or a party, even in the absence of immediate material benefits, due to their confidence in the potential future utility of personal bonds to individuals in position of power. And, in fact, these relationships and worldviews have, so to say, stood the test of time.

Finally, going beyond the classic approaches to these issues, centered on institutional views of public space, participation, and citizenship, our intent was to understand how individuals "resist" this segregation, what kind of strategies they develop to live in and "use" the public space, and how they experience forms of social mobility that oppose common understandings of their situation modeled on stereotypical and demeaning formulations like the "culture of poverty" model.

Notes

1. Some of these areas, especially those that were encompassed by the city center, which in the past were considered marginal for being too internal, are today perceived as central (Di Leo 1985, p. 46), while other areas, such as the southern neighborhoods of the Zona Sud, continue to be perceived as peripheral in spite of the fact that they are between five and eight kilometers from the city center.
2. Usually, the main perpendicular axes that connect the hills and the sea retain the name of the stream over which they were built. The most important are *Torrente* San Filippo, *Torrente* Gazzi (South), *Torrente* Zaera (Center). But in this case, today's term for that neighborhood, Viale Europa, has prevailed; *Torrente* Annunziata and *Torrente* Giostra (North).

3. The organized struggles of these communities in places of work and in the matter of housing have been discussed in Farinella and Irrera (2014) and Saitta (2018).
4. On this paradigmatic disaster in contemporary Italy, on its management and the effects in terms of speculation, corruption, political instrumentality, and demagogy, see v. Özerdem and Rufini (2013); Contreras et al. (2017).
5. What was taking place in Messina during the 1920s was also happening elsewhere, for example, in Rome and Milan. During the 1920s, the main political problem for Fascism was, in fact, the "absolute necessity of including building property in the new order of national economic forces" (Bortolotti 1978, p. 53)—even though this meant stifling the social base represented by the small- and medium-sized property owners who had supported the rise of Fascism in favor of larger building companies, which were now given new opportunities for integrating rent and profit (Bortolotti 1978, p. 67). If in cities such as Rome, where there had been no earthquake, the inner areas were demolished and "gutted" (*sventrate* is the word Fascism had used), in places that had been almost destroyed, such as the area of the Strait, where nature had done most of the work, there was all the more reason to move forward using the same criteria. Likewise, if, like in Rome, those working-class suburbs called *borgate* were built to contain the homeless masses that had been produced artificially by the national government (Ferrarotti 1970), in Messina, the same authorities were building low-income housing and neighborhoods for people evicted by natural events.

References

Abu-Lughod, L. (1990). The Romance of Resistance: Tracing Transformations of Power Through Bedouin Women. *American Ethnologist, 17*(1), 41–55.

Barrat, C., & Green, G. (2017). Making a House in Multiple Occupation a Home: Using Visual Ethnography to Explore Issues of Identity and Well Being in the Experience of Creating a Home Amongst HMO Tenants. *Sociological Research Online, 22*(1), 9. Retrieved from http://www.socresonline.org.uk/22/1/9.html.

Bayat, A. (1997). *Street Politics. Poor People's Movements in Iran.* New York: Columbia University Press.

Bortolotti, L. (1978). *Storia della politica edilizia in Italia.* Roma: Editori Riuniti.

Caporale, A. (2010). *Terremoti Spa. Dall'Irpinia all'Aquila così i politici sfruttano le disgrazie e dividono il paese.* Milano: Rizzoli.

Castles, F. G., & Ferrera, M. (1996). Home Ownership and the Welfare State: Is Southern Europe Different? *South European Politics and Society, 1*(2), 163–185.

Chubb, J. (1982). *Patronage, Power, and Poverty in Southern Italy.* Cambridge: Cambridge University Press.

Comune di Messina. (2012). *Regolamento comunale "Criteri per l'assegnazione e la gestione degli alloggi di edilizia residenziale pubblica"*, approvato con Deliberazione del Consiglio Comunale n. 31/C – 19.04.2012. Messina: Comune di Messina. Retrieved from http://www.comune.messina.it/il-comune/regolamenti-comunali/documenti/regolamento-comune-di-messina-approvato-con-deiberazione-n-31c-del-19-04.pdf.

Comune di Messina – Dipartimento Politiche per la casa. (2018, August 23). *Relazione al Sindaco sullo stato del risanamento.* Messina: Comune di Messina.

Contreras, D., Forino, G., & Blaschke, T. (2017). Measuring the Progress of a Recovery Process After an Earthquake: The Case of L'Aquila, Italy. *International Journal of Disasters Risk Reduction, 28*, 450–464.

D'Amico, L. (2012, March 28). Messina: La mafia catanese aveva messo le mani sul cantiere di Fondo Basile. L'occupazione degli alloggi: gran parte degli abusivi risulta essere residente in altre zone! *Gazzetta del Sud*, ed. Messina.

D'Amico, L. (2017, November 19). Fondo Fucile e quei "belvedere" del degrado. *Gazzetta del Sud.* Retrieved from http://www.gazzettadelsud.it/news/messina/267389/fondo-fucile-e-quei-belvedere-del-degrado.html.

De Martino, E. (1973). *Il mondo magico. Prologomeni a una storia del magismo.* Torino: Einaudi.

Desmond, M. (2017). *Evicted. Poverty and Profit in the American City.* Allen Lane: Penguin.

Di Leo, L. G. (1985). Stratificazione urbana dal secondo dopoguerra e recupero dell'esistente. In G. Di Leo & M. L. Curzio (Eds.), *Messina, una città ricostruita. Materiali per lo studio di una realtà urbana* (pp. 45–48). Bari: Edizioni Dedalo.

Falconieri, I. (2017). *Smottamenti. Disastri, politiche pubbliche e cambiamento sociale in un comune siciliano.* Roma: Cisu.

Farinella, D., & Irrera, O. (2014). Eterotopie della resistenza e classi subalterne. Infrapolitica e mobilitazione per il lavoro in un'azienda sanitaria del Mezzogiorno. *Etnografia e ricerca qualitativa, 2*, 195–218.

Ferrarotti, F. (1970). *Roma da capitale a periferia.* Roma and Bari: Laterza.

Foucault, M. (1978). *History of Sexuality* (Vol. 1). New York: Random House.

Foucault, M. (1993). Of Other Spaces: Utopias and Heterotopias. In J. Ockman (Ed.), *Architecture-Culture 1943–1968: A Document Anthology* (pp. 420–426). New York: Columbia University Graduate School and Rizzoli.

Foucault, M. (2007). *Security, Territory, Population.* New York: Picador.

Frank, K. A., & Stevens, Q. (2007). Tying Down Loose Space. In K. A. Frank, & Q. Stevens (Eds.), *Loose Space. Possibility and Diversity in Urban Life* (pp. 1–33). New York and London: Routledge.

Giddens, A. (1990). *The Consequences of Modernity.* Stanford: Stanford University Press.

Hollander, J. A., & Einwohner, R. L. (2004). Conceptualizing Resistance. *Sociological Forum, 19*(4), 533–554.

Hyra, D. S. (2017). *Race, Class and Politics in the Cappuccino City*. Chicago: The University of Chicago Press.

IACP (Istituto Autonomo Case Popolari). (2018). *Riscontro a nota 209728 del 23 Agosto 2018.*

Ipsale, M. (2015, May 4). Alloggi di via Taormina, gli occupanti non vogliono andare via. Parte l'informativa in Procura. *Tempostretto*. Retrieved from https://www.tempostretto.it/news/tensioni-alloggi-via-taormina-occupanti-non-vogliono-andare-via-parte-informativa-procura.html.

Jiménez-Domìnguez, B. (2007). Urban Appropriation and Loose Spaces in the Guadalajara Cityscape. In K. A. Franck & Q. Stevens (Eds.), *Loose Space. Possibility and Diversity in Urban Life* (pp. 174–192). New York and London: Routledge.

Lefebvre, H. (1991). *The Production of Space*. Malden: Blackwell.

Lewis, O. (1966). The Culture of Poverty. *Scientific American, 215*(4), 19–25.

Lo Verso, A. (2015, March 17). La storia delle case popolari. Colpa della burocrazia. *LiveSicilia*.

Ortner, S. (1995). Resistance and the Problem of Ethnographic Refusal. *Comparative Studies in Society and History, 37*(1), 173–193.

Özerdem, A., & Rufini, G. (2013). L'Aquila's Reconstruction Challenges: Has Italy Learned from Its Previous Earthquake Disasters? *Disasters, 37*(1), 119–143.

Pruiti, R. (2010). *La città negata. Viaggio nelle baraccopoli di Messina a un secolo dal terremoto*. Roma: L'Espresso.

Regione Sicilia. (2012). *Legge Regionale n.4/2002. Interventi per l'accelerazione e il completamento del risanamento della città di Messina*.

Saitta, P. (2018). *Prendere le case. Fantasmi del sindacalismo in una città ribelle*. Verona: Ombre Corte.

Scott, J. C. (1985). *Weapons of the Weak. Everyday Forms of Peasant Resistance*. New Haven, CT: Yale University Press.

Scott, J. C. (1990). *Domination and the Arts of Resistance: Hidden Transcripts*. New Haven, CT: Yale University Press.

Serra R. (2017, November 3). Quartiere Taormina, l'ex baraccopoli trasformata in discarica. *Giornale di Sicilia*.

Sik, E. (1988). Reciprocal Exchange of Labor in Hungary. In R. Pahl (Ed.), *On Work: Historical, Comparative and Theoretical Approaches* (pp. 527–547). New York: Basil Blackwell.

Spain, D. (1993). Gendered Spaces and Women's Status. *Sociological Theory, 11*(2), 137–151.

Stavrides, S. (2007). Heterotopias and the Experience of Porous Urban Space. In K. A. Franck & Q. Stevens (Eds.), *Loose Space. Possibility and Diversity in Urban Life* (pp. 174–192). New York and London: Routledge.

Stevenson, D. (2003). *Cities and Urban Cultures*. Maidenhead and Philadelphia: Open University Press.

Stornante, F. (2013, March 16). Fondo Basile: alloggi pronti da due anni ma si vive ancora in baracca. *Tempostretto*. Retrieved from https://www.tempostretto.it/news/risanamento-fallito-fondo-basile-alloggi-pronti-due-anni-vive-ancora-baracca.html.

Tempostretto. (2008, September 14). Baracche San Matteo. Lo chiamano risanamento.... *Tempostretto*. Retrieved from https://www.tempostretto.it/news/spazio-quartieri-baracche-san-matteo-chiamano-risanamento.html.

Thompson, S. (1994). Suburbs of Opportunity. The Power of Home for Migrant Women. In K. Gibson & S. Watson (Eds.), *Metropolis Now: Planning and the Urban in Contemporary Australia* (pp. 33–45). Leichhardt: Pluto.

Tumino, A. (2015, June 6). Case basse di Paradiso. 19 occupazioni abusive. *Gazzetta del Sud*, ed. Messina.

Tumino, A. (2016, December 1). Giostra, otto milioni sprecati dal 2007. *Gazzetta del Sud*, ed. Messina.

Zampieri, P. (2018). *Esplorazioni urbane. Urban art, patrimoni culturali e beni comuni*. Bologna: Il Mulino.

CHAPTER 7

Messina Today: Representation, Identity, and Mobilization for Change

7.1 The Shacks as a "Performative" Space: The Subalternization of the Slum-Dwellers from an Institutional Point of View

The previous chapter focused on the centrality of the earthquake and, more particularly, on its visible legacy: the lasting presence of the slums. It is also the central element around which local identities are constructed, in the constant tension between representation, stereotype, and historical reality. In this respect, discourse about Messina often produces a stereotypical and essentialist image, focusing on the slums, which, in turn, express backwardness, underdevelopment, and barbarism. In short, this discourse expresses a modernity that is out of balance, the causes of which can be traced to specific cultural traits that can be simplified into the incompetence and corruption of the political-administrative system, and the culture of the slums and slum-dwellers or both.

However, as Herzfeld reminds us (2005, p. 15), "within a political structure defining the shape of cultural identity, people are constantly and ineluctably drawn into binary choices. Nation-state ideologies tend to divide the world into Manichaean pairs and to coerce or seduce their citizens into adopting the same rhetoric for the moral organization of their own everyday social relations. But people's uses of that rhetoric may be deliberately irreverent or even subversive".

D. Farinella, P. Saitta, *The Endless Reconstruction and Modern Disasters*,
https://doi.org/10.1007/978-3-030-19361-4_7

Starting from an analysis of the discourse, identitarian representations, and practices, this chapter aims to show that stereotypical and essentialist identitarian representations can sometimes be changed, adapted, and politically activated in order to make specific claims and can also, often be reversed.

As Palumbo (2001, p. 120) suggests, "the behaviors, practices and symbols used in power relations in order to stereotype and classify the 'other' and place them in a position of subalternity, can also be used by the weaker party as a tool to subvert the same power relation. [....] The analysis of resistance strategies, of compromise, acceptance or imposition, must be expressed in the reading of the social poetics performed, from time to time, by real actors".

Identitarian representations are part of a discourse that implies that power relations remain open, however uneven they may be. According to Foucault (1978, p. 93), "power is not an institution, and not a structure; neither is it a certain strength we are endowed with; it is the name that one attributes to a complex strategical situation in a particular society". If then, "where there is power, there is resistance, and yet, or rather consequently, this resistance is never in a position of exteriority in relation to power", should it be said that "one is always 'inside' power, there is no 'escaping' it, there is no absolute outside where it is concerned, because one is subject to the law in any case? Or that, history being the ruse of reason, power is the ruse of history, always emerging the winner? This would be to misunderstand the strictly relational character of power relationships. Their existence depends on a multiplicity of points of resistance: the strictly relational character of power relationships: these play the role of adversary, target, support, or handle in power relations" (ibid.). Such points of resistance are "present everywhere in the power network. Hence there is no single locus of great Refusal, no soul of revolt, source of all rebellions, or pure law of the revolutionary" (Foucault 1978, p. 95). Instead, there is a "plurality of resistances, each of them a special case: resistances that are possible, necessary, improbable; others that are spontaneous, savage, solitary, concerted, rampant, or violent; still others that are quick to compromise, interested, or sacrificial; by definition, they can only exist in the strategic field of power relations" (Foucault 1978, pp. 95–96).

The aim of this chapter is not only to analyze the objectivation strategies and tactics put in place by the *outsiders* (those who do not live in the slums), in line with their role (journalists, politicians, strangers, or the citizens of Messina), but also the tactics used by the slum residents who, at times, perform exactly as expected of them, while exploiting some of the stereotypical representations that have formed over the years, either overturning or abandoning them, but always within the realms of power relations.

First, it is important to see the codified narrative that describes the city from an external point of view, whereby the slums are "earthquake slums" that have existed "since the earthquake". In this narrative, the people living in today's slums are directly descended from the first slum-dwellers and have spent whole generations waiting for a house that never seems to materialize. This sort of interpretation implies an ideological premise: in a modern Western capitalist society, the very existence of these slums is a scandal, and, therefore, given the exceptional nature of the situation, the responsibility must lie with the territory itself. This form of cultural blame therefore becomes part of the discourse used to refer to the local population and local institutions.

Messina: a hundred years of living in slums. This is the emphatic title of a special article published by the Corriere della Sera (a leading national newspaper), in 2008, in which the journalist Cesare Fiumi sarcastically commemorates the hundredth anniversary of the earthquake, stating that it is the historical legacy of the city:

> The Messina bridge[1] spanning this part of the city is an affront to the eye and an offense to reason. It is a wooden plank about three meters long, designed by poverty to cross the slurry-filled 'straits', an open-air sewer for drainage and urine, which separates two rows of slums. It is a single rotten span put in place by those who have either the courage or necessity to live here, in houses that cannot be defined as houses, with asbestos roof coverings and internal walls crisscrossed with cracks and ceilings covered in mold. These are the Favelas of the Giostra district. And this is not even the worst area of the city that has made its slums a monument to decay, with its *one hundred years of slumminess*: ever since the early morning of December 28th 1908, when a devastating earthquake [...] swept everything away, every building. [...] Because these favelas in Giostra—and those in Camaro and Fondo Fucile—[...] are not part of a new immigration crisis, but the terrifying homes of Italians: the third generation of slums, the grandchildren of the first shacks donated and put up by the Swedish and Americans, the Swiss and Prussians, the day after the earthquake. [...], before the brick slums built by the fascists [...], and, after the bombardments during the last war, and the Republic came up with the idea for this "temporary public housing". (Fiumi 2008)

Since the beginning, the article introduces the reader to the stereotypical features that make up an institutional account of the Messina slums. In this dramatic and emphatic narrative, the inexorable "slumminess" of the entire city takes shape, where the slums exist as a unique and contradictory moral-historical legacy. Such notion is elaborated further over the following lines, and mixed with the "human element":

> [...] Generations of slums and generations of "Messinesi" who have lived and continue to live there, over 3,000 of them in the year 2008, and 100 years after the earthquake, in areas including the Annunziata, the Fondo De Pasquale or Giostra, like the geological layers of Italian history, of the Sicilian political class and not, and its squalor. Today, the Messina slums are a lesson in favela-style open-air architecture, where mafia infiltrations and water infiltrations through open gashes in the ceiling are one and the same thing, [...] the Signora Concetta's shack is like being in the trenches, on the front of human catastrophe: one single room, a toilet hidden behind a cardboard door that can do nothing to hide the smell, and the outside sink. Beyond, a cement courtyard where rats walk freely: "I killed one last night, by hitting it on the head with a spoon". Her only spoon.
>
> Signora Concetta is eighty years old and has lived in this shack her whole life. It is one of the original ones from 1909, financed by the 30 million lire allocated by the then prime minister Giovanni Giolitti. "First my parents died, then my brother just up and left and I never saw him again. I worked as a house cleaner, and I was good and fast. I was just waiting for someone to give me a house. It's been promised so many times, but I'm still here". (Fiumi 2008)

Within this typical narrative structure, Concetta—a character that journalists love, as is shown later—is the element that incarnates the connection between stereotyped representations of history and the human with its quotidian.

> She talks with difficulty and walks around barefoot among the rats because her shoes are so uncomfortable on her swollen feet. But she continually smooths down her grey, sticky hair and in the end, between tears of exhaustion and an exaggerated laugh to help force back the tears, she opens one of the plastic bags that serves as a cupboard and shows me a photo of herself, when she was young and pretty. And the wooden shack, like the future held no fear. [...] It shows Concetta Albano and the shack that was erected for her while the mafia were killing Joe Petrosino in Palermo, it seems like the perfect metaphor for the "infinite earthquake"—for false promises, wasted

aid, for the termporary that becomes permanent—and a an utterly exhausted city, region, nation.

[...] In these corners of Messina, that the city is so ashamed of, it is as if the earthquake and war had only just come to an end. They don't like to talk about it, except when it's time to vote, and the word goes round that everyone will get a house soon. And [...] the evictions, the demolitions and the new constructions are delayed because the hand brake is on as far as bureaucracy is concerned [...] Out of sheer desperation, in the end, the families become life-long earthquake victims. Once they are allocated a council house, they leave the slum in the favelas to their children, in a never-ending chain of poverty (Fiumi 2008).

Soon in the article, Concetta gets to embody her community and becomes a collective subject—"They". "They", then, form a precariousness, which, by its very persistence becomes permanent, and the expression of a double failure. The first being represented, clearly, by a political-administrative system unable to guarantee the rights of citizens. These "favelas" are even more "scandalous" since they exist in the heart of modern Western "civilization". This is not about a "distant" other, someone who is undeniably "different" from us (the immigrants, as suggested in the same article), nor are they the expression of a temporary need; on the contrary, they speak to us about another "neighbor", who has always been one of "us" (i.e., part of Italy, Europe, the West) and who gives us a discordant, misshapen view of ourselves, as if we were looking into a distorting mirror.

But for those who have lived in this place unfit for human habitation, it is the only inheritance they have to pass on, the only treasure they have to offer [...] "There's no hope here". Here, during the winter of 1909, a little over a month after the earthqake, when the city was still a pile of rubble, but also gave the impression of being a building site, Luigi Barzini of the Corriere *expressed* the hope—it was the 4th February 1909—that a "great future was being prepared for Messina". But it didn't last long, when the help that had been sent from all over the world was sent home, and the illusion embarked along with them. Already, on the 9th May the *shacks*—a word that has lasted for over a century and is still part of the narrative—were in the headlines: "during the allocation of the shacks, and against illegality and favoritism, the police had to shoot into the crowd, leaving 5 dead on the ground." Since that time, on the ground, the shacks are still part of Messina: they have survived the King and Fascism, two world wars and 61 different governments of the Republic. They are monuments to a century of Italian history. (Fiumi 2008)

Implicit in this part of the discourse, buried among the lines, lies the second failure. The failure of the slum-dwellers themselves, who, from generation to generation can do nothing but remain, unable to escape the poverty (exacerbated by their precarious living conditions) or to escape the slum culture, while remaining attached to the only "treasure" in their possession, that can be handed down from father to son.

At first glance, the article seems to side with the slum-dwellers, presenting and celebrating them in this story about the "oppressed". The final part clearly reiterates this idea and mentions an episode involving a police shootout that took place during the struggle for the allocation of a shack, and which resulted in five deaths. This image reinforces the idea of the slum-dwellers as victims of the State, relegated to the margins of society, and obliged to remain there because there is no alternative (public housing), all of which fuels the persistence of the slums (by building temporary and social housing). Thus, the slum-dwellers are the victims of the false promises made by politicians, who only show interest in their plight at election time when they can expect a vote in exchange for the promise of something (the slum-dwellers are unable to refuse while perfectly aware of the hypocrisy). There are also promises of financial aid and funds that the administrators, either through ineptness, negligence, or convenience, lose or "misplace". However, while the article follows and enriches the life stories of the slum-dwellers, they also appear guilty of allowing themselves to remain in this situation; they are victims of their own inertia, suspended in the precariousness of eternal waiting, and have become passive and immobile. As such, they seem to have regressed to an infantile state of dependency, in need of a house, and, more generally, in need of someone who is in a position to provide for all their needs, precisely because "a hundred years of slumminess" have left their mark. To cite the famous novel by Márquez, the reference here is to a hundred years of solitude, backwardness, and isolation, passed on from one generation to the next (to paraphrase the title of Gabriel Garcia Márquez's novel referred to in the journalist's article). This is the result of the rather ungenerous image of Signora Concetta, whose extreme poverty is witnessed through some of the topical references: the plastic bag that functions as a cupboard, the single spoon used to kill a rat (and, one supposes, to eat), her one-room dwelling, where the toilet is hidden behind a cardboard door, and her barefoot shuffle through the rubbish because she is "ill" (her feet are swollen), and above all, brutalized by life, with her solitude presented in the image of a brother who has never returned. There is also the bitterly observed photo of herself as a young woman. Here, the journalist has taken one single case and absolutized it, hypostasized it as an emblem of life in the slums. Obviously, what we know about Signora Concetta does not correspond to reality (even though the minutely described details have a neorealistic, almost documentarist ring),[2] but this is the story the journalist wishes to tell. In his narrative, the extreme poverty experienced by Signora Concetta is not part of life's trajectory, or a path of progressive existential precariousness; on the contrary, it is represented as a fixed, immobile, objective state, inherent to living in a shack: the incarnation, therefore, of an inescapable destiny. Poverty, dependence, need, and inertia become, in a simplified and reductive way, the very essence of the shacks themselves, and the effect is the total subalternization of the slum-dweller, whose personal story, biographical trajectories, aspirations, motivations, and choices are completely nullified. Whoever lives in a shack does so forever; they are necessarily poor and in a state of need and, above all, trapped by passivity, unable to imagine an alternative characterized by self-determination, but constantly waiting for someone to give them a house, a job, something. It is precisely at this point that a certain ambivalence toward the slum-dwellers starts to make itself felt, as the tendency is to describe them as victims, but also as guilty ones, since they are incapable of autonomy. The veiled reference to an alleged exchange vote, which is as widespread as it is useless, has the same effect, especially when we consider that the shanty residents are still there, generations later. It seems they cannot resist this type of vote, as if they were naively obtuse, and devoid of any logic or critical abilities (the same sort of moralistic deduction made in many discussions regarding the "poor").[3] Two further ambivalent situations demonstrate that the shanty-dwellers are responsible for their own situation. One is the simplification of the widespread phenomenon of buying and selling the shacks (implying an opportunistic calculation of costs and benefits as part of complex family strategies), which romantically become part of "an infinite chain of poverty", linking the members of the same family. The family perceives the "shack" as the only precious asset they have to pass on, as an asset that has been accumulated with great difficulty in the midst of the permanent poverty they inhabit. They are proud to pass it on to relatives who are still in need, sanctioning their inability to leave the "shanty culture", since they are always ready to relegate other family members. Finally, there is an ambiguous reference to mafia infiltrations, with the inevitable gang members hiding inside the slums, just like the infiltrations of water coming through the ceiling. In the article, this analogy seems, once again, to show the inescapable destiny of the slum-dwellers; not only are they poor, but they are probably criminals too, unless, of course, they have managed, heroically, to resist; like the slum-dwellers who mop up with whatever they can get their hands on when the rain comes through the roof. Besides, the shanty residents, as the journalist states emphatically a few lines earlier, "work (even when there isn't any work)".

This article was chosen here because it gives a good summary of the institutional perspective from an external point of view (national and local press). The shacks represent a scandalous exception for Western modernity, poverty, and need are the essence of the people living there, and there is passivity and dependence, the mafia and criminality, the exchange vote and the exploitation of the slum-dwellers for political patronage by the political-administrative system. The shacks represent people's individual dwelling places and their active lives (people are born, live, and die there), and they are also a family legacy (passed down from generation to generation) so that their inhabitants are unable to abandon the slum culture they have internalized.

This essentialist discourse helps to establish exactly what it means to be a slum-dweller and how they have become the object of what Foucault (1988, p. 51) called "practices of subjectivity"; since they have "always" lived in conditions of poverty, they are denied a personal history and a subjectivity, and are objectified and transformed into one of the neutral needy. In this way, they have become subalternized.

However, it is not just the individual who is subalternized; it is also the collective space. To say that the Messina slums are "the earthquake slums" and have existed "since" the earthquake, but that they are essentially the same, despite a few obvious changes and rearrangements, makes it impossible to construct a shared story of how the slums have grown over time. The majority of those dating back to the earthquake have been demolished or completely transformed (therefore, in a contradictory and ambivalent way, they are both the same as, and different from, before), while other shacks have been added, enlarging an already existing area, multiplying and materializing in areas where they previously did not exist. They have become part of working-class neighborhoods, tacked onto social housing, making it increasingly difficult to distinguish between post-earthquake shacks and subsequent ones, between illegally built houses, self-builds, and social housing, creating a continuity and a circularity (the one produces the other and vice versa) that is underlined in Chap. 4.

If we consider the shacks an exception, then we deny that they are intrinsically part of the poorer districts and their growth trajectory, and the specific outcome of policies (and non-policies, or rather, the lack of public intervention) that exert the spatial effects of categorization and governance on the subjects living there. Public action that moves the lower classes to the margins of the city produces disciplined spaces and subjects by using criteria based on legal but arbitrary rationales, and defines, classifies the lower classes, and within this group the poorer classes who, all things considered, might be content to live in a shack or temporary housing. This inscribes the geography of the inequalities of the individual body in a collective space, thus creating the "subjugation of the body through space" and expressing the isotopy of the disciplinary apparatus (Foucault 2006, p. 52 passim) and permitting the constitution of a "schematic and centralized individuality" (Foucault 2006, p. 50).

However, the shacks have not disappeared, nor are they a conspicuous exception in the areas in which they are located; similarly, the slum-dwellers are not caged animals, confined to the margins and isolated from the rest of the neighborhood; on the contrary, they are intrinsic to and an expression of it. In fact, the shacks seem to fit in and belong with the other public housing, and are the demonstration of how these areas have come together because of unresolved tensions between:

- public actors who institute urban confinement in order to normalize, classify, control, and subordinate;
- private businesses which, in the chaos left by the lack of legal territorial planning, especially in the more marginal areas, try to create profit opportunities, even if this means accentuating disorder and environmental degradation;
- the communities living in these de-personalized, de-personalizing, and anomic areas, to which, despite everything, they need to give meaning, since this is the space in which they experience life.

On the one hand, these peripheral areas in Messina are emblematic of many other peripheral areas around the world and are the unexpected outcome of public action and the market. In essence, they are a "residential reservoir" (Bellicini 1990, p. 40) and the expression of the utter inability to plan a social, shared, and collective space, and also, the denial that it might be possible.

The juxtaposition of fragmented habitations, indifferently thrown together, highlights the absence of a shared architectural language (Bellicini 1989, p. 38). This peripheral area of the city is a "hybrid", contradictory place, an alienated borderland on the margins between the city and the country. It is a no-man's land, scarred by the untidy presence of abandoned farmland interspersed with houses, illegal structures, and the ruins of old farm buildings, electricity pylons, and undefined patches of land (gated, fenced off, illegal garages, small cultivated gardens etc.) which, like a junkyard, becomes the repository for all the things that cannot be seen in the official city (Bellicini 1989, pp. 43–47).

On the other hand, in response to these mechanisms of depersonalization and objectivation, many of the practices put in place by the subjects living there can be read as forms of subjectivation and personalization of space, as we suggest in Chap. 5.

There are many examples of the ways in which the residents of these areas re-attribute sense to the space, always with heterotopic effects, and producing, "different spaces", "other places", that are a "mythic and real contestation of the space in which we live" (Foucault 2009, p. 25).

One example is the common practice of erecting a commemorative plaque on the street where a loved one or a "member" of the local community has lost their life. It begins with a photograph attached to a traffic light, or a few plastic flowers, and slowly evolves, growing over time until some of them become stone or cement monuments, with a vase for fresh flowers, and poems or delicate hand-written phrases stuck on with tape and covered with plastic to keep them dry. Other examples are the numerous votive shrines dedicated to the Saints or the Madonna, which spring up inside the courtyards of social housing or in niches built into the walls of self-build houses, either as religious symbols or as expressions of faith, or the long gateways and fences made of repurposed scrap material (such as old bedsprings), or flowerbeds transformed into private vegetable patches, and again, motorway ramps that become a circuit for clandestine horse races at night (to mention just some of the cases encountered while carrying out this research). In all these cases, the actions break the uniform and predictable interpretation of space, and generate a displacement effect and a topographical shift. For example, a gravestone (which is not actually a gravestone because no one is buried there) transforms an anonymous street—a place of transit and objectively of no interest to passers-by—into a stopping place, a meeting place, and a place of personal memories. In short, in the re-appropriation of public spaces which have become personalized and privatized, heterotopias arise, becoming "counter sites" (Foucault 2009, p. 24), "areas located in the general space of our experience, that assume a certain material reality, but also reveal how the space is experienced, challenging its ordinary use" (Sabot 2012, p. 24), making forms of social and self-experimentation possible and permitting, therefore, opportunities for subjectivation.

Perhaps it is within this frame that we should read the areas of informal housing dotted around these poor neighborhoods, which redefine the whole concept of the slum, with reference to the need to personalize spaces and social needs according to specific requirements that are often simply ignored by the institutions.

As Fera and Ginatempo (1982, 1985) had already revealed in their research from the 1980s, much of the irregular self-building in the South emerges as an adaptive strategy, rather than as an alternative, and as a way of reconciling the dominant values of society (aspirations to house ownership and the middle class) with local problems (scarcity of economic resources, precarious work, the lack of low-cost housing on the market, and long waiting lists for subsidized public housing). In this context, self-building highlights the way in which complex family networks are activated, including the mutual exchange of labor, economic resources, and social resources (Cremaschi 1990, p. 140), in what is a traditionally European, spontaneous type of self-building. Ward (2002, 2004) saw it as a characteristic of rural economies, dictated by the implicit recognition of the right to own a house, the informal strategies put in place (in other words, parental support and family help with building), and "borderline" practices.

The most interesting aspect concerns the capacity for self-building to autonomously meet or reflect expectations that are often frustrated by heteronomous housing policies, which tend to impose a top-down and paternalistic vision of social needs. As Cremaschi (1990, pp. 147–148) points out, from the end of the Second World War, national building policies remained trapped between two coexisting, almost schizophrenic representations: on the one hand, and especially in the South, the solution to the housing shortage and overcrowding of the 1950s was to expand the quantity of building stock, by any means, even by sectoral support, and a chaotic overlapping of benefits, exemptions, and privileges which supported the residential building boom.[4] On the other hand, there was the idea, which was a particularly Left wing point of view, that housing rights had been socially polarized, with some of the population socially guaranteed, while others were permanently marginalized, so that it was important to reclaim control of housing production for better social redistribution. Stemming from this was the idea that the housing emergency was strictly related to the phenomenon of irregular building and the building of slums (most prominently in the case of the Roman suburbs. See: Ferrarotti 1970) within a triangle that included poverty, social need, and the imperative to modernize policies. The overall response, as far as housing policies were concerned (an increase in the housing stock, support for ownership, a rise in social building), reflected a heteronomous reading of housing needs that was applied "in relation to standard figures, or to local particularities, to the possibility of obtaining resources, to the original expectations of the population, and the chance that they might be taken into consideration" (Cremaschi 1990, pp. 147–148). This administrative definition of housing needs, which homologates and is indifferent to local demand and its transformation on a practical level, reinforces social exclusion (which in abstract, the policies aim to contrast) and legitimates the sort of informal/irregular practices which transform space and creates ways to sidestep the rigidity of such policies.

This type of "do-it-yourself" building, which characterizes many of the poorer class neighborhoods in Messina, manages to respond in an adaptive and flexible way to the needs of a population that is mobile and constantly changing and where the family unit has the ability to increase or decrease according to necessity. Far from being trapped in the slums, the subjects and their families circulate between the various housing opportunities in the neighborhood, in a sort of family DIY, in which every single opportunity is left open, and the precariousness of the housing situation goes hand in hand with informal work, with different forms of self-production, self-consumption, and family self-help (see Chap. 5).

Research undertaken in areas of urban regeneration shows that there are few cases of immobility (i.e., those who were born into and have never left the slums); on the contrary, in some areas, it is more common to find young couples with children (as a recent visit to the slums confirmed). This data is also in line with statistics reported by the Municipality of Messina (2018), which recently observed, based on the experience of previous rankings, that around 40% of subjects currently recorded in the areas of regeneration may not have the necessary prerequisites, stipulated by law L.5/2002, for the allocation of housing (in other words, continuous residence in a shack between the years 1995 and 1998. See Chap. 6 on this point). This may be the reason why young couples, who are often perfectly aware that they are paradoxically "irregular" in their irregularity, make the choice to live in a shack temporarily, as a way of coping with the critical phase and expense of starting a new family, even if it is not always easy to leave as quickly as they would like. On the contrary, they are sometimes obliged to return (for various reasons, which may include job loss or separation). If so, the empty, deserted shack may be depressing and in need of refurbishment, but it represents a way of economizing and of helping the subject to cope with unexpected issues. In this way, and to help with the family economy, the shack may be handed on from one subject to another, or rented out, sublet, and so on.

As already mentioned, the peripheral area is an open, unfinished, borderland place and, for this reason, it is both de-personalizing and individualizing. The lack of any form of shared collective space implies that individuals will need to resolve their own problems and construct their own spaces, adapting them to their own necessities. In comparison to the general idea of "unfinished", which can seem so depressing, the idea of an unfinished house (whether it be a shack or house) which can be adapted, enlarged, or altered to deal with the changes taking place in the subject's life, becomes a sort of active response to the personalization of the neighborhood. Thus, even when allocated through public housing schemes, the house inevitably opens itself up to informal practices, becoming a potentially open construction, an ongoing building product that can be adapted to new solutions (Bellicini 1989), be enlarged and changed to suit the transformations taking place in the family living there, while taking into account their economic possibilities and relational needs.

7.2 The "Shanty Tour": Come in, Come and See the Conditions We're Living in!

"Come in, come and see the conditions we're living in!", a woman tells a journalist as they make their way into the alleyways of Fondo Fucile, in what is just another flying visit to the area to report on the scandal surrounding the "infinite slums". Here is the news story that places poverty on center stage: they are familiar words—the same ones heard many times over as we wandered around the neighborhoods during our research.

It is a theatrical performance, or if we prefer, one of "political identity" and "social poetics",[5] played out each time there is contact with the institutions (the local and national mass media, public officials, politicians): the poverty, the deterioration, the filth, and the conditions "we are forced to live in" are what the journalists expect to report, and this is what the "Messinesi" slum-dwellers, who have been "living in the slums since the earthquake 100 years ago", put on show, so as not to disappoint expectations. The accusations have been made.

> *News Report broadcast by TG2 (a leading news channel) on 8-8-2018—Fondo Fucile Shantytown*
> *Narrator: Messina, Fondo Fucile Shantytown*
> [The cameraman enters the alleyways where the sounds of daily life can be heard, pots, pans and dishes being moved around, and a woman shouting in the background "Oy, can you delete that shot, eh??" A man accompanies the journalist, followed by a small procession of kids. He stops outside a house and grinning, shouts out in Sicilian dialect, as if to emphasize his role as part of the community: "Luciaaaa….. chiddri di casiiiii…." [Luciaaaa…..It's those people about the houses] to warn her that the television people are coming. Then in a lower tone, with the mocking expression of one who thinks they know it all, he says, "come on, Lucia, let us into the house for a look."
> Journalist—Can I come in?
> Lucia—Please, come in, come in, enjoy the view (the woman is now speaking with a softer voice and in Italian)
> The narrator continues with growing emphasis—Outside it's almost 40°C, it's even hotter inside. It's difficult to breathe, with your breath rising up to the asbestos ceiling (there is a shot of the ceiling), poison!
> Interview question: Lucia, how old are you?
> Lucia—53!
> Journalist—And have you always lived here? (emphasizing "always")
> Lucia—Nods her head and then answers: 31 years (our observation: by our calculation, she went to live in the shack at the age of 22, probably straight after getting married, with her new family).
> In the background the voice of the same woman who was shouting at the beginning can be heard, but she is even angrier and has come into Signora Lucia's house where she begins shouting from another room "semu cunsumati e ruvinati se sdurupunu i casi ca dintra!" [it'll be a disaster for us if they demolish the houses]. One of Lucia's sons is sitting on a plastic chair fanning himself, and on a small table next to him is a photo of some children in a silver frame, probably her grandchildren. The camera operator shakily films the asbestos ceiling.
> Journalist—Have you bought the house? (Meanwhile, the camera operator films a close-up of the kitchen units. Next to these are some small pieces of furniture. The camera gives a rapid view of the various rather kitsch ornaments in the room: silver photo frames, a small table full of tiny figurines, a bouquet of plastic flowers)

Lucia—Oh yes!
Journalist—How much was it?
Lucia—(Lucia hesitates) Eh, eh, well ehm.... In lire, a million lire (about 500 euro—a very small sum of money).
Journalist—A million lire, for this house! (The journalist sounds scandalized, as if to say no one would want a house like this even if it were free).
The narrator continues, while the camera films a group of houses: "this nightmare of sheet metal has been here since 1908, when the earthquake killed half the population. The survivors were crammed in here. A settlement that was supposed to be temporary but is still here. With 2000 inhabitants! In Sicily, 11% of families live in conditions of absolute poverty. This shantytown in Fondo Fucile is emblematic of this hardship!"
The camera continues to film the streets of the ghetto, showing children playing on their bikes, until it frames a group of young men all around 20 years old.
The journalist stops and asks them a question, but the editing only shows one boy's answer: "it's full of asbestos! And it's so bad, asbestos causes cancer!"
The next shot shows washing hanging out to dry on a balcony, a symbol of life in the South, and then the corrugated rooftops.
The narrator continues—"And yet a regional law dating from 1990 allocated 500 billion lire to build *real* homes for the descendants of the earthquake victims, and to demolish this city within a city" (obviously regardless of Signora Lucia's incoherent testimony, in which she had candidly confirmed to have bought the shack, and not to have inherited it from her family). "Now, the newly elected Mayor has the difficult job of trying to make sense of this shameful situation". (We move to the Town Hall, to the new Mayor's office)
Journalist—Where did the 500 billion lire go?
Mayor—Well, part of the money is still being held by the social housing institute, so now we're going to ask for this money to be issued, because our idea is that by the end of the year, that is, December 2018, we need to find 2000 existing houses to rent, so we can move these people into the houses and immediately demolish the slums.
Journalist—So, in just 6 months, you'll be able to resolve a problem that has been going on for 110 years?
Mayor—That is our plan and that is what this agency is for. (The scene returns to Fondo Fucile).
Journalist—The Mayor has said that in six months, the problem will be resolved, and they'll take you away from here.
The boy from before answers—"yeeees, we know! We believe the Mayor!" (His tone clearly shows he believes they are being taken for a ride!)

> Another boy drinking coca cola moves away saying—"Iavi 'na vita… ora ni dugnunu i casi!!!!" [we've been waiting a lifetime… they wanna give us houses now?] (As if to say it's impossible!)
> Journalist—Don't you believe him? (smiling)
> The boy shakes his head.
> Another excerpt; a different boy appears, though he doesn't want his face to be seen on film, so stays behind a wall: "If you move us out of here into rented accommodation but you don't give us a job, what's the point if in the end we end up with no house, I mean, come on!"
> In the background, we can hear that the statement about getting a house, but also the fact that they will be moved, has created some confusion. The woman who had been shouting angrily is now quite worried about the fact that they might be moved. In dialect, she says: "E picchi n'avissuru mannari i ca, a mia non mi 'ni 'ntocca casa!" [Why do they want to send us away from here? I'm not eligible for social housing!].
> Journalist—If they come to take you away, do you want to stay here?
> The boy explains—That's not it! It would be better if they gave us a job and then a house!
> Narrator—here, the fear is that this is just another unfulfilled promise, and so it's better to keep this shack, in the end without a job, without any idea of what the future holds—this is the only certainty.

This news report is very interesting because the stereotypical and essentialist discourse regarding the slums referred to in the previous paragraph has been reproduced in full—this time in visual form. Signora Lucia candidly confesses that she has lived in that house for thirty years, probably following her marriage and the formation of a new family (she is fifty-three years old). In the same way as other witnesses in our study (see Chap. 5), we can assume that in response to the urgent need for housing and in the face of low economic resources, she pragmatically decided to buy the shack from someone who was leaving (because they had been assigned a house, had bought a house, or simply moved to another city). Despite this, the narrator mystifies the reality of the situation, at least on a moral level, by repeating that the slum-dwellers have lived there since the earthquake, and that today's residents are their living descendants. The narrator goes on to establish a correlation between living in the slums and poverty (Fondo Fucile is emblematic of the hardship experienced by 11% of the families in Sicily who are classified as poor) and emphasizes the poverty and deterioration there by means of the "tour", first inside the house, where they focus on the ceiling, lingering on the knick-knacks in the kitchen which become the folkloric, kitsch features that highlight the "slumminess" of the people. After a brief institutional pause, during which the journalist pays a visit to the mayor, the video shows that a reasonable and civilized solution does exist: to demolish the slums and transfer the residents to rented housing. We then see how the slum-dwellers receive this idea and their ambivalent attitude, which shows them as victims but also guilty of historical complicity in the clientelistic game of false promises. Their disjointed claims only underline their helplessness and dependence on the welfare system, in their insistence that they should be "given" not only a house but also a job. It seems they are unable to leave the slum culture, which is also the culture of poverty. It is the same culture that recognizes the slum as the only asset they possess and therefore as something to be preserved. The video focuses on certain aspects in order to emphasize the narrative: the crude poorly made windows and doors that stand out in the narrow streets; the discarded debris and filth in the streets; the washing hanging between the houses; the noise of everyday life; the reference to the rats "dancing" among the houses; and, obviously, the image of children playing in the filthy alleyways, proof of the injustice of a situation that also affects these innocent lives.[6]

However, on closer observation of the video, the inconsistencies start to surface, small discrepancies, which show that the slum-dwellers who appear in the film are actually perfectly aware of the game of "representation" they are playing, and that they have chosen to tell the journalist precisely what he wants to hear. The inhabitants display their "slumminess" with a certain nonchalance. This becomes clearer when the man accompanying the journalist to Lucia's house stands outside and shouts loudly and exaggeratedly to announce the journalist's arrival, while quietly suggesting that she gives the journalist what he wants to hear during the tour of the house. Finally, we see it in their astounding refusal to leave the slums. With a half-smile, the young man tries to explain that they are not refusing to be housed; the problem is that if they move into rented accommodation, but remain unemployed, the risk is that they will become homeless (if, e.g., the funds for their rent run out). In addition, aware that the video will create a stir, he strategically exploits the opportunity to test the ground, by sending out an outrageously frank message (of which he is perfectly conscious—in fact, he asks not to be filmed) that is also extremely effective: we want a "house" *and* a "job".

Throughout the video, and especially in some of the comments in Sicilian dialect that can be heard in the background, but were not eliminated during the editing process, as they lend "color" to the story even though they are incomprehensible to a non-Sicilian audience (the report is being filmed for a national channel), it emerges that the residents' greatest worry is linked to the fact that many of them do not have the necessary prerequisites for the allocation of accommodation. Without these, they are afraid they will be labeled as "cheats", or "illegal by profession", as, indeed, one Messina resident defines them in a post on the mayor's web page. Thus, even if their right to accommodation is not formally recognized, at least they can demonstrate their need in a material way, by remaining faithful to their shack, which, in its "squalor" demonstrates their need and material poverty, rendering them "deserving" to external eyes, regardless of the prerequisites (see Sect. 7.4).

The video also calls to mind a situation that took place in 2001, while one of the authors of this book was delivering Census questionnaires in a part of the city near *Torrente* Gazzi (Case Garufi—Via Gaetano Alessi), an area behind the city prison and one of the most infamous in the city.

As you enter via Gaetano Alessi, the first building you see is a large villa, once the property of a noble family, consisting of several floors and surrounded by a fence and a large plot of land. The villa is in a state of crumbling abandon, or at least this was the researcher's impression as she approached in search of a doorbell or intercom. On hearing a voice, she realized that there were a series of small buildings to her right, inside the courtyard. A young woman of about thirty invited her in. She was carrying a small child while others peered out of the house, and after greeting and enquiring as to her visitor's identity, her immediate impression was that the researcher was a representative from the local institutions, in this case, the "Municipality of Messina". "Come in, come and see what conditions we're living in!" Although she was there to distribute census questionnaires, the researcher attempted to explain that she was not a council official ready to assign the young woman a house. However, in the eyes of the young woman, she was the embodiment of a possible connection to the institutions. The young woman therefore began her performance, acting out her part as she moved through the various rooms in the "house", which seemed more like a badly equipped garage with a huge steel roll-up door at the entrance. A dim light revealed a few small rooms, some without windows: the walls were obviously damp and there was a badly installed sewage pipe running the length of the ceiling down to a poky bathroom. "There, see, can you smell that terrible stink?" the woman said. The hardship of her situation became obvious during the tour of the house, and she wanted it to be witnessed. However, she was beginning to realize that the researcher was not there in an official capacity, perhaps she was not even a municipal official, and thus, the possibility of her interceding on behalf of this poor family was very slim. Yet, she was there with her questionnaire, ready to record information for the census, and this was enough to link her to the institutions, as someone who could potentially testify on behalf of the young woman, according to the logic of "you never know". After all, making another "complaint" costs nothing; it is simply a repeat performance.

By using the terms "repeat performance" and "acting", the intention is not to question the family's obvious hardship, nor to imply that it is untrue. Objectively speaking, this young woman and the residents of Fondo Fucile seen in the television news story live in slums, in illegal housing riddled with damp and mold, with asbestos roofing and countless other problems.

Nevertheless, some of the issues raised (the presence of asbestos, damp, mold, and garbage) are generally the same problems that affect most poor neighborhoods. For example, the presence of asbestos in schools built during the 1970s, or in the individual water tanks installed in all social housing (not equipped with an autoclave system) or on the rooftops of blocks of flats. Much of this asbestos is still in situ, and the old asbestos water tanks have still not been replaced by new polyethylene ones. Sometimes, it is possible to see well-maintained shacks regularly repainted by the owners so that mold cannot form, as well as legal IACP houses that are completely rundown. In many cases, the poor quality of the materials used and the construction work carried out make social housing particularly vulnerable and in need of constant maintenance in order to avoid internal and external deterioration, almost as much as the shacks themselves. This is the case of the Arcobaleno Houses mentioned previously.

From this point of view, the regeneration of the slum areas would seem to be an important step in the right direction. However, it would not resolve the problem of the more general deterioration of the poorer areas, which require a longer-term restoration plan that recognizes the wider picture and no longer thinks of the slums as a permanent emergency. Indeed, the elimination of the older areas of slums and shacks would not resolve the situation because, in the meantime, thousands of others have sprung up in many of the peripheral areas. These are illegal self-builds that are erected within the space of a few nights, enlarged and refinished over time, or abandoned buildings that are reoccupied. Alternatively, they are ghost houses (unregistered), only known to the authorities because they have incurred penalties and sanctions. The municipal police then seal these and give out eviction notices.

7.3 The Right to Be Here: Making the Invisible Visible

The area of *Torrente* di San Filippo, which is not a post-earthquake construction area and has been recognized, together with some other areas, as being in need of regeneration by the regional law n.10/1990, is a good example of this type of situation.

Walking around these buildings (the illegal self-builds inside courtyards, private vegetable gardens, illegally built garages, conservatories that occupy sections of the street but which are tolerated because they were built by members of the community) the resulting disorder brings to mind a heterotopia of spaces that have been repossessed, reutilized and redefined according to the needs of the residents. As already mentioned in Chap. 6, the material presence of this do-it-yourself patchwork of buildings seems to emphatically affirm its own right to exist and be part of the local landscape (not just a passive symbol of need defined by others) and its right to give meaning and sense to space. However, the material nature of these buildings is in direct contrast with the abstract, formal, and bureaucratic rationality of the public apparatus, which dictates that something exists only after it has been formally recognized, certified, and classified, according to previously defined criteria and requirements imposed from earlier (on the basis of an interpretation of need which, in order to be impartial, can only be decided by the State) and, therefore, presented as objective. Thus, the actual presence of these constructions is not enough to guarantee social recognition, which needs to be established and certified by the public apparatus. Only in this way can the invisible become visible and make sure their rights are recognized. Without certification, these ghost families remain in their ghost houses. Paradoxically, it is with this general objective in mind that the slum-dweller—the illegal house-owner, the subject who is in a position of marginalization with regard to the institutional guarantees represented by the welfare system—activates, utilizes, and adapts the stereotypical identitarian representation of themselves to their own needs.

Perhaps this is the reason why the slum-dwellers are eager to act out a part during the "slum tour", why they give a representation of themselves, which, in the collective imaginary, constitutes the frame of poverty and extreme hardship. In the representation in which the subject says: "come in, look at the conditions I'm living in", they are playing with this stereotype. At the same time, they are not expressing the reality of their situation—nor giving an objective interpretation of the facts in a chronological order of succession—but performing the most politically effective of narratives on the institutional, public, and media stage. "Making the invisible visible" means exaggerating the poverty and hardship of living in a shack, the unsanitary conditions (the open sewers and rats that "play" with the children, the "slumminess"), to the point of objectifying the need and rights to a house, regardless of whether or not they possess the legal requirements.

Nevertheless, "slumminess" is not always a frame that works. It can be decidedly counterproductive in the case of illegal self-builds that do not fall within the so-called requalification areas but are built in other parts of the popular districts (in fact, they are numerous, dotted all over the city, often hidden away and barely visible from the outside, known only to the other residents who are "part" of the neighborhood). In all these cases, the emphasis is on the positive role of the subject, on how their new construction is an improvement to the area; what is required of the institutional interlocutor is that they simply register the building, in order to make the invisible visible. To illustrate this point, two different accounts follow (already briefly mentioned in Chap. 5).

The restored house: Wandering through the area, the researcher notices that, close to the IACP courtyard, there is a fence extending out toward the road. The whole fence has been covered in a fine green mesh to cover up the self-built house hidden behind it. The researcher rings the bell at the entrance gate, and a young man opens. He too is convinced that his visitor is possibly a public official, and he invites her in. Over coffee, the young man proudly shows the researcher around the house he has built himself. There is a sitting room with kitchenette, furnished with a large red and dark wood wall unit, with spotlights for directional lighting, all in the most up-to-date style. The bathroom has been particularly well styled—there is a large corner bath with a tiled decoration of an oriental woman. The observer is struck by the care that has been taken over the smallest detail and, in a certain sense, finds confirmation of the hypothesis that self-built houses more precisely reflect the needs and requirements of their builders than houses bought on the regular housing market. The young man is married with a small child, and he explains how he built the house by reclaiming a part of the street that was being used as a general rubbish tip, where the locals dumped old tires and other garbage. As he tells his story, he emphasizes the amount of work he put in to reclaim and improve the land: a veritable restoration of a blighted area. He proudly points out all the elements that make his a real house rather than a shack.

And then he gets to the crux of the matter: he tells the researcher all about his failed attempts to move his family's "official residence" to this house, saying that the Council refuses to recognize it as a house and therefore does not recognize the young man and his family, forcing them to maintain separate "places of residence" (each with their respective family of origin). It is the paradox of his life, and there is apparently no solution. It is not possible to register the invisible, but it is important to continue asking someone to take responsibility for it because only by registering the invisible can you make it effectively visible.

The conservatory: There is a similar situation in of the Case Arcobaleno. When the observer arrives on the scene, the owner shows her where to park, in one of the many internal alleys that have become cul-de-sacs; parts of the street that have been slowly incorporated into the houses. The owner invites his unexpected guest in. His wife has recently died, and he is alone with a disabled son and other older children who live in nearby apartments. The house is well looked after, with a classical style that imitates the middle-class interiors of the 1980s. The first room is the sitting room, where a huge collectors' doll dressed in flamboyant evening clothes is seated in the center of the room. Further on is the classic glass-fronted display case, which holds all the "valuable" objects, including the little ornaments (*bomboniere*) received on special occasions. Then, there is the classic English-style sofa in pastel colors. We go into the kitchen to get a pen. Here, the solid wood kitchen is immaculate. Generally, the family eats together in the conservatory, and, since his wife's death, it is where his daughter cooks. We enter, and see his children and grandchildren are all waiting to eat. He explains how he has worked hard over the years to build the conservatory himself. It is made of glass and cement, and there is a large built-in kitchen (again, constructed by the owner), with a big electric oven for external use from which they have just taken some baked focaccia, and a big wooden table in the center of the room. The Council has brought a lawsuit against him and would like to demolish the conservatory as it illegally occupies part of a municipal road. On the other hand, he argues that the self-build is legitimate because it does not create any difficulties for others (the closure of this secondary road does not obstruct free passageway), and he has realized a life-long dream. Once again, by showing the beauty of his self-built structure, its functionality, and symbolic value (for family gatherings), he is asking for the invisible to be registered, to make it visible.

7.4 The Messina Residents Observe the Slum-Dwellers: Who Is Truly Deserving?

As already mentioned, in the classic interpretation of the narrative, the shacks and shack-dwellers interest local politicians only because they constitute a natural reservoir of votes during elections, fueled by false promises, as witnessed by the continuing existence of the shacks.

However, this is a profoundly reductive vision. As mentioned in Chap. 5, from a particularistic-clientelistic point of view, a house is often seen as one of the potential benefits provided by the welfare system, and which can be accessed by adjusting the pre-requisites. Even when these are officially defined, they can appear arbitrary, variable, subjective, and fragmentary because they are based on categories or particularistic characteristics that are the result of pressure from various social groups.

For the subject who sees them as a possible way of accessing the institutional guarantee system, which would allow them to gain social mobility, the requirements and benefits that derive from a particularistic-clientelistic model are clearly not universal. The game is therefore played openly, and the subjects who play it are perfectly aware that the opportunity to enter the protection system, and the necessary prerequisites for doing so, is limited (they can only be accessed through possession of the correct credentials), temporary (they might be suspended at any moment or might change due to new social pressures), and elastic (if the categories are vague and based on particularistic criteria defined ad hoc but are unstable and potentially changeable, even the requirements can be forced and stretched by those subjects who use specific strategies in order to qualify, ranging from formal possession of the requirements, acquired dishonestly, to informal strategies).

Paradoxically, for the marginal subject, the whole experience of interaction with the welfare and social guarantee system is based on the logic of the *deserving* and *undeserving poor* (Bridges 2017). Individuals have access to welfare only if they are "deserving" of it and can demonstrate their state of "worthiness" by possessing the correct prerequisites, which are determined at regular intervals. In this way, the subject wishing to enter the protection system is exposed to a disciplinary power, which defines the requirements necessary for access, and which reveals its isotopic character that both individualizes and depersonalizes the subject. In fact, this power appears to be arbitrary and discretional, and devoid of what, for Weber (1958), was the main source of the legitimacy of rational-legal power, that is, compliance with the rules according to the criteria of impartiality, objectivity, and universality.

However, since power is always a situation that implies both strategies and gaps, regulations always contain a way out, opportunities for action that should not be interpreted romantically as resistance, but as tactics for gaining recognition and inclusion; once again, *to make the invisible visible.*[7]

In fact, the various practices described in Chap. 6 and in this chapter—for example, buying a shack, transferring residence to a "different" family unit so as to qualify for a place on the rehabilitation waiting list, building a shack in an area due for rehabilitation, or, again, illegally occupying as yet undesignated social housing, playing the "vote in exchange" card, and so on—are all actions with one aim in mind: that of gaining the right to recognition and access.

Far from being isolated practices, albeit in a different form, they also concern the middle and upper-middle classes, as has already been pointed out in relation to the question of unregulated building (see Chap. 4). However, while middle-class citizens are generally quite tolerant of informal opportunistic practices that allow them to obtain some advantage or benefit they are not entitled to, they morally condemn the lower classes, in particular, the slum-dwellers, describing them as "parasites". This can be seen in a selection of the negative comments left on the social networking sites that accompanied the national news story in August 2018, analyzed in Sect. 7.2:

> Woman A—110 years after the earthquake struck Messina and Reggio Calabria, they're demanding the right to social housing (probably left to them by their great-grandparents or perhaps even their great great-grandparents!!!) well, honestly? Those same people who declare they haven't got enough money to buy or rent a house, and who go around in cars worth 30 thousand euro, or scooters worth 10,000 euro, and they have smartphones they bought for 1000 euro. I reckon the stupid ones are my parents who've been paying a mortgage for 30 years.
> Man A—it's not worth their while to leave: *cu barraccotu nasci barraccotu voli stari!* [Those born into the slums want to stay in the slums!] Not only do they want a decent house but they want the council to give them a job! Well, you have to go and look for work like everybody else who wants to live a decent life, but this doesn't apply to them. Stay there then, and live like pariahs! Because if they don't, then they have to pay taxes and so they have to work. It's not worth their while.
> Man B—100 years, are you kidding me? They've been given thousands of apartments—the parents leave and their children take over. Let's just stop

talking about these earthquake slums. It was all political cronyism anyway! I've been married for 40 years, and for 20 years I've paid rent and now I'm paying a mortgage, with all the money I've paid to GESCAL [the Fund for the Management of Workers houses, destined for the building and allocation of workers' houses]. Instead of giving the houses they built to those who paid, like me, and as they'd done in other cities, they were given to the slum-dwellers. So let's not talk about the earthquake, you know lots of them don't even want to pay the rent, they've got two-storey villas, the shacks are empty, others don't really stand a chance, the houses should be given to these people.
Man C—It's true. These people could emigrate to the North or go abroad to start a new life, but no, they prefer to be professional spongers; they don't pay a penny of taxes or for services. Instead, they go around with expensive cars and motorbikes, they use the latest cellphones, and they also want the Mayor to give them a job. If they don't start to get them out of those shacks, their children and grandchildren are going to be living there for the next 50–100 years!
Man D—Well, they're like the Roma, you know water, electricity etc. etc. all free, it's difficult to get them out! I remember these shacks already existed in 1966, so they've earned the right to them by usucaption! 😀

It is obvious from these examples that in the local collective imagination, the slums and slum-dwellers do not possess the more stereotypical features of poverty or social need that are typical of "external" accounts of the city, and there is little concern that they have been in the shacks since the earthquake. On the contrary, in the comments, there is more emphasis on the shacks as products of political clientelism and opportunism, and the shack-dwellers as parasites of the welfare system. Their condition of need appears to be false, as witnessed by their ostentatious and conspicuous consumption of luxury goods, and their involvement in informal-illegal activities is also implied. The belief is that together with political clientelism, the buying and selling of the shacks has produced a multitude of underserving usurpers who have infiltrated the slums and who should be distinguished from more deserving citizens. From here, it is a short step to the abstract logic of guilt and the concept of the deserving/underserving poor. The difficulty of defining objective criteria or of finding real solutions to the problem means that the generic request for police involvement or repression (such as removing the rights to housing for whoever has served a jail sentence or has been under house arrest) gives the idea that these subjects are on parole. They can achieve this status only if they can demonstrate, somehow, that they are deserving, or if they agree to disappear.

> Man E—Those who put forward the idea that today's shacks are the direct consequence of the earthquake is not being honest or doesn't understand the phenomenon. Many of these shacks are recent and were built a long time after the earthquake. Some of them were built to meet a need for housing, others weren't. Before stating that we need to find housing for everybody, we need to verify who has the right to it and who doesn't. Actually, before making any announcements they should do some research, with all the technology on offer, for example an aerial survey with drones, to prevent the usual wise-guys from building extensions or building new shacks so they can trade them for real housing.
> Woman B—Well, they could start by removing the people who don't have the right to social housing, but were assigned these houses because they have "connections". In Messina, there are lots of empty houses, but for years they have been assigned to people who have a normal salary, and a stable job. Also houses taken away from those who really need them. It would be a good starting point. But I'm not saying anything that isn't already known.
> Man E—There is a way of avoiding the re-occupation of the empty shacks; by immediately cutting off the electricity and/or services, and the immediate demolition of the slums. With a metal perimeter fence surrounding the whole area and surveillance. Then, immediately start building the new housing units destined for those who are effectively in need and have the right.

In this climate of increasing distrust toward the good faith of the shack-dwellers, they have no other alternative but to play the game, subjecting themselves to this humiliating form of public "means-testing", by once again putting their need on show (including the classic slum themes of rats, damp, sickness and poverty), thus confirming, not just their right to a house (because that is defined by law, according to a set of prerequisites) but also the social legitimacy of their request. These ideas are very clear in this exchange between Messina residents, published on the new mayor's social networking site, where he announced the imminent clearance of the shacks:

> Woman A—In reality, many people have been assigned housing over the years BUT THE SHACKS HAVE NEVER BEEN DEMOLISHED and are therefore always reoccupied by other people.
> Man A—The shacks are passed down from generation to generation or sold to the highest bidder. There are people who can't afford to buy a house or worse, that don't want to, because they're waiting for social housing. It all needs to be demolished with no ifs or buts, and kept under surveillance.

Woman B—And let's not forget the Fondo Saccà requalification Area. It's just a pity that we can't post photos, otherwise I'd show everyone the disgraceful conditions my husband and I live in with our two daughters aged 7 and 10. [...]
I've tried, but I can't publish the photos.
Dear Mr Mayor, I live in a shack with my two daughters, the eldest of whom has been diagnosed by the ASL as having health problems. When it rains I have to get the buckets out and we have maggots in the house... I still remember catching a huge rat, as big as my foot. You certainly need a lot of courage to sleep with a beast like that in the house. Once a year I have to call the pest-controller to gas the bathroom drains. And the amount of rat droppings I've collected... Mr. Mayor please help us.
Man B—It's all just propaganda... the shacks have always existed and they always will... because it suits everyone. Some healthy local and national policies would help prevent the continuation of this deterioration, build some decent social housing and prevent the empty spaces from being reoccupied....
Man C—Let me just remind you that the people who have been living in these houses for years are not poor, because if they own more than one motorbike or scooter, then they can certainly pay the rent even if they are doing undeclared work, but they don't, because they expect to be given a house and so they prefer to live in these conditions.
Woman C—Dear Sir, however, you can't tar everyone with the same brush!
Woman B—Exactly. I haven't even got a car... And I can't afford to rent a house either. If I could, I would have left ages ago.
Woman D—Dear madam, I really hope for your sake that the Mayor's project is carried out as soon as possible!!!
Woman E—Few families here manage to support themselves and it's true that they sell the houses but we've never managed to discover hot water. Maybe it's worth it for some of them in the hope that they'll eventually get social housing so in the meantime they don't pay for rubbish collection and water?
Woman D—Believe me, there are many families unable to get by! And as for the rubbish and water, those who obey the law pay!!!
Woman E—They should be torn down, but more importantly, there should be a waiting list based on need. In my opinion, Woman B should be given a house immediately.

7.5 From Salvini to De Luca: The Political Party of the Slums

According to Palumbo (2001, p. 130) "the imagined South risks becoming not just a stereotypical, essentialist representation, a naturalized and more or less 'resistant' identity claim, but a stereotypical, essentialist, naturalized model, for politically manipulated identity claims". And we can analyze the visit of Matteo Salvini, national leader of the League Party (and currently Deputy Prime Minister and Minister of the Interior) from this point of view. On November 2, 2017, he visited the Rione Taormina in Messina for a personal "tour of the slums" in view of the 2018 political elections.

At this point, a little background information is necessary. At the time, the popularity of the League party was in decline due to a series of scandals linked to the misuse of funds, concerning both the founder of the party, Umberto Bossi, as well as some regional governing bodies (e.g., the Region of Lombardy). The new secretary, Matteo Salvini, therefore, decided to create a new path of action for the party, and his election campaign was almost entirely focused on his own personal skills as a pragmatic man of action, who was closer to the people than to the corridors of power. He wanted to perform the "average" person and be someone capable of understanding and giving back to the "people", using the same simple, direct language as the man or woman on the street, without any apparent filters or concessions to political correctness. It was a cleverly constructed image created mostly on social networking sites, where Salvini has kept up an artificial but constant and direct relationship with the "Italians", especially on his very popular Social Network pages. Salvini had also realized that it was important to broaden the League's electoral base, which was historically in the North of Italy, to include the South. He had therefore symbolically changed the name of the party from Northern League to League, forming alliances with many local independentist parties in the South of Italy. Finally, he had changed the symbolic contents of the party and its original identity, which had been created during the 1990s in the wake of strong anti-meridionalism, and from the primitive slogan "The North first", had changed to the nationalistic "Italians first". It was a change of slogan that implied a strong anti-immigrant agenda. Immigrants were presented as the main cause of the country's economic difficulties and alleged social deterioration, and once again, we can see how stereotypical and essentialized identitarian constructions can be re-adapted and re-functionalized according to new organizational needs. This strategy has led to the League becoming the third most voted party and to an unexpected post-vote coalition with the *Movimento 5 Stelle* (Five Star Movement), to form a hybrid government that is strongly populist, anti-European, reactionary, and with xenophobic traits that are apparent in its constant fight against immigration as Italy's main problem.[8]

In the video news story posted online by the newspaper journalist for Voce di Popolo, Salvini agreed to give an interview, having posted the customary photos of the slums on his Twitter account (with a photo of a dead rat on the grass as gallery opener) after his visit there.

Before summarizing and analyzing the interview, it might be a good idea to explain that the Rione Taormina is part of the Gazzi area designated for regeneration and is an emblematic expression of the paradox of the unfinished restoration, which needs, forgive the pun, to undergo some regeneration itself. A road cuts across the Rione, dividing the north side from the south. Briefly, during the second half of the 1980s, the northern part had already been cleared of slums to make way for new social housing, the construction contract going to the then president of the city's soccer team, a property developer with strong interests in the area (which already concerned the old city football stadium) and with suspected mafia connections.[9] Once finished, the building began to deteriorate very quickly due to the low quality of the materials used, and today, it is in a state of total decay. The slum clearances continued in the remaining areas on the north side, but in the cleared areas, the building of the new flats progressed very slowly. Twenty-five years later, in 2015, a new IACP complex was completed, next to the now dilapidated building erected by the former president of the soccer team, and which was destined for the families who were highest on the Council's housing list, especially those from the Fondo Fucile area (Redazione 2015). However, even before the accommodation was allocated, it was illegally occupied by some of the families from the south side who were lower down on the waiting list (Redazione 2015). A few months later, in 2016, the Council managed to clear the new housing and allocate it to the people who were officially entitled. At the same time, the clearance of the south side began, although it was rapidly interrupted after only two months due to lack of funds, leaving some of the houses earmarked for demolition still standing (some of which were illegally reoccupied). Moreover, the municipality and the IACP played a continual game of "pass the buck" as to who was responsible for the removal of the rubble, cement, and materials from the demolition of the cleared houses.

As Bottari commented (2017) in a leading local newspaper, "it has become the embarrassment of the south side, the disgrace that makes all passers-by wrinkle their noses, which day after day, after 21 endless months of slum clearance, since March of last year, has seen the weeds turn into trees, the garbage become overgrown by brambles, the debris visible at every turn, and rats scuttling around freely amongst the ruins". This is the situation on the day of Salvini's visit. As the news story opens, we see Salvini standing right at the center of the neighborhood (recognizable by the dilapidated houses in the background) surrounded by some of the residents (who every now and again come closer to shake his hand or kiss him on the cheek) and by journalists.

> *(A woman's voice in the background—Well done Salvini, if you're here it means you understand the situation!).*
> Journalist—Honorable member Salvini, excuse me, can you describe the situation you have seen here today?
> Salvini—Well, you can see for yourselves. It's incredible… I asked, when was the last time a Mayor came here, or a council member … zero! Zero! When you have kids who just want to go out for a game of football, mothers who are simply asking for water to come out of the tap. When people talk about rats, roofing, broken glass, wood…
> Journalist—Have you seen the conditions inside these houses?
> Salvini—Eh now, we'll see… I'm not here to do miracles, but commitment and seriousness, yes. It seems absurd that these administrators have left the situation like this… not just for a year … for a whole lifetime (little smile)… It's not acceptable, this is 2017, we're in Italy. Well, we're in Messina, but we're in Italy, we're not on the other side of the world, you know!
> Journalist—It's a very serious situation here in Sicily, cavaliere Berlusconi said so just the other day, it's tangible!
> Salvini—Yes, of course, but all you need is a little normality! You just need a little honesty here, competence, a little money, miracles aren't necessary. In fact, they asked me about "the bridge?"… the bridge… perfect, okay, but in the meantime, the people need a little light, give them some water, give them a roof, a little grass. What's really missing is the ability to do something normal! We don't need big promises. You could restore normality here in just a month.
> Journalist—What's missing are the everyday needs?
> Salvini—Yes! What's missing is a bit of intelligence! How can you talk about major projects when there are situations like this! [New Report Broadcast by Voce di popolo.it, 2 November 2017, Salvini's visit in view of the regional elections]

If we analyze the inferences in the text, we can see that Salvini is emphasizing the contrast between the need for "normality" (where children can be children, or play a game of football), and the deterioration, summed up by his references to "roofing, broken glass, rats", which implicitly refer to his previous tour of the slums and alleyways (not shown on the video). The situation is unacceptable and scandalous, both because it has been going on for so long ("a lifetime!") and because we are in Messina, which is still a part of Italy. Moreover, it is the year 2017, this is not the Middle Ages, nor is it the other (non-civilized) side of the world, by which he obviously means this is not Africa or the Third World. In fact, with just two or three comments, his discourse brings to mind the simplified and stereotypical slogan, "Italians First", which proved to be such a winner, going straight to the heart of the Italian electorate. Finally, the search for someone to blame, and the instrumental use of local politicians and officials, who, with their inability to act, mixed with their dishonesty, reinforced by the (unverifiable for those listening) affirmation that nobody, neither "mayor" nor "council member", has visited the area recently (perhaps only during election campaigns seeking votes). Here too, the inference is clear: vote for me.

However, Salvini is not alone in making strategic or tactical use of stereotypical representations. A similar discourse strategy was used by the new Mayor Cateno De Luca, who was elected in June 2018, during elections that saw an almost 70% abstention rate. Immediately after his election, in an unexpected and surprising move, he decided to try to gain legitimacy and consent for his political party by dwelling on the idea of the slums and shacks.

The mayor had also made several tours of the city slums during his election campaign, but nobody expected that he would begin his mandate by immediately tackling the problem, which, in the past, had been subject to much more protracted political timing. The new mayor is considered an outsider because he is not originally from Messina but from a nearby town, and a little "provincial". He managed to win the elections to everybody's surprise, with a campaign based very much on his individual abilities, for example, his professed skills as a rapid and pragmatic administrator and decision-maker. Nevertheless, he was unable to get any of his own party elected as counselors and now finds himself in a vulnerable position, without an overall majority in the Council where his decisions need to be approved.

Just two months after taking office, the mayor announced his wish to go ahead with the allocation of houses for families residing in the requalification areas by October 31, 2018, and by December 31, 2018, with the demolition of the remaining slums and clearance of all the resulting detritus. The first proposal to resolve the problem was renting, and then the purchase of existing houses from private owners. In order to carry out these requalification operations, he set up a new ad hoc agency and put forward a request to the National Government for the recognition of a state of emergency, linked to the deterioration of health and hygiene conditions in the slums, so as to obtain the appropriate financing. He also asked for clarification regarding the funds linked to the Regional Law n.10/1990, as well as to the state of social housing and the real number of families currently living in those areas. He also strategically shifted the question of blame implicit in some of the official discourse surrounding the slum-dwellers, to the issue of health: the slums should be eliminated as quickly as possible because the presence of asbestos was creating a health hazard, increasing the risk of tumors and other respiratory diseases. Cleverly, he was not attributing blame (what came first, requalification or the slums? Do the slum-dwellers create clientelism or vice versa?) but acting to protect the health of the population and to resolve a verifiable health emergency.

In the wake of an international trend, he set up a direct line of communication with the citizens of the city, periodically using social networking sites where he would publish his daily work programs, photos of meetings, debates on proposals, and any other notice or document that could testify to his hard work, accompanied by declarations, considerations, accusations, and jokey comments, all contributing to the image of a mayor of the people, a decision-maker, who loves his city and puts respect for the rules at the center of his mission. He is either known as the "sheriff" mayor or Cateno lo "Scatenato" (the Unleashed)—a play on words which he used as the mantra for his electoral campaign. Here is an example of this communication strategy in relation to the question of the shacks, taken from the Mayor's Social Network page:

> "EITHER THEY GET RID OF THE SHACKS OR THEY GET RID OF THE MAYOR OF MESSINA!"
>
> With this statement, made yesterday evening at 8.00 pm in the Prefecture, my meeting with the Deputy Prefect, XXX, came to an end, after a heated debate over the draft for council orders concerning the clearance and demolition of the shacks and slums.

"Excellency, I cannot be left alone to manage over one hundred years of hypocrisy and mismanagement. The National Government must acknowledge and declare a state of housing emergency with the appropriate emergency measures and funding, otherwise I am ready to exercise my prerogative as the highest local authority in matters of health and hygiene and civil protection".

I have no idea what will happen to me, but having analyzed almost 30 years of talk since the famous law 10/90, I am convinced that the regional and state institutions must face up to the cold hard facts.

Let's just wait and see who tries to stop an irreversible situation that exposes delicate and dangerous scenarios.

I AM THE MAYOR!

Attached is the order signed tonight, that will be published on the official noticeboard.

By 31st October 2018 everybody will be out of the shacks

By 31st December 2018 the demolition and disposal of all relative materials will be completed

It is still early days to understand whether his strategy of putting the rehabilitation problem top of the agenda so quickly and diligently, which is certainly unique in the history of the city, will be more than just a promise, confirming the political hypocrisy mentioned by Brunson (1993), or whether it will really lead to an acceleration of the clearance process. What appears evident, rather, is the new mayor's choice to legitimize himself on the back of his own policies for rehabilitation, having understood the symbolic value for the citizens and, consequently, his potential return in terms of social consensus, as being the only one able to strengthen his political position, even though he doesn't have a political majority. As some citizens enthusiastically comment on social networking sites:

Man A—If you manage to get rid of the shacks, you will go down in the history of the city as the greatest Mayor of all time. The Number one 💪
Man B—WE'RE MAKING HISTORY here IN MESSINA
Woman A—I've been hoping this would happen for years ... Now I feel very moved ... come on Mayor... dignity to this city, the gateway to Sicily!
Woman B—More and more... it can't be true 😀 this Mayor is giving us hope!!! Let's pray x Cateno who needs our prayers! Cateno is the best—he's unique □🏆🎖🏆

This symbolic inauguration by the city has already proved to be an important move toward the approval of the Rehabilitation Agency by the Council at the beginning of September 2018, even though some council members have expressed misgivings about the new agency's proposal, and the Council seems to want to delay or even reject the measure. The threat of the mayor's early resignation expressed in the phrase "get rid of the shacks or get rid of the Mayor of Messina" and communicated on social networking sites to the citizens of Messina, has, in fact, forced the Council to approve the project, under penalty of seeming overly involved in specific obscure interests, clientelistic, and indifferent to the city's real needs.

In conclusion, although it is difficult to pass judgment on the plans from a political and even historical point of view, as far as tactics are concerned, we can say that the rapid acceleration of the rehabilitation plan by the newly elected mayor is important on three levels:

- *Symbolic*: with the strengthening of social consensus, the citizens will at least partially protect it in the face of the Council's political majority.
- *Political*: because it helps to strengthen his political legitimacy on a regional and national level, presenting him as a decision-maker, an active man who aims to end what has always been seen as an infinite rehabilitation plan, within a specific deadline. This will take place as part of a program, which, in the mayor's words, sees the regional Sicilian government as its next political objective.
- *Economic*: because reopening the rehabilitation scheme and linking it to the request for a state of public health emergency would potentially allow him to obtain new funds and financing (which is what one would expect of a southern mayor, after all). It would also allow the sort of speculative economic mechanism that brings the building sector and public funding and contracts together (previously interrupted due to lack of funds), which are so typical of the city. In addition, the funding and renting or purchase of thousands of empty houses/apartments, at a time when the housing market is in recession, would help redistribute resources and increase the number of supporters, not just from an ideological point of view but also from an interest point of view. The creation, that is, of a real "social bloc".

An update. In mid-October 2018, the national government refused to recognize the state of emergency, effectively blocking the possibility of obtaining special and urgent funding for the realization of urban rehabilitation. Without these fundamental public resources, the new mayor's battle for the clearance of the slums came to a complete halt. Up to October 31, 2018, and contrary to what had been promised, the shacks and slums located within the areas of the historic rehabilitation plans have not been eliminated. The allocation of public housing to the slum-dwellers continues slowly, according to plans and projects that had already been organized by the previous mayor.

Contrary to the first few months of his mandate, the new mayor now avoids posting news and official communications about the rehabilitation process on his personal social networking pages, preferring to allow the question to fall into oblivion. These latest developments seem to suggest that his political strategies were probably based on economic questions and the possibility of mobilizing economic resources and resuming the dynamics of a speculative economy, which have always characterized the city. The national government's rejection of the state of emergency has blocked this possibility (as additional emergency funds cannot be requested). Consequently, the subject of the slums and slum-dwellers has once again been relegated and has had to take a back seat.

Notes

1. This provocative reference is to the elusive bridge across the Messina Strait, which would connect mainland Italy with Sicily, and which, from the Second World War onwards, has been seen as a strategic and symbolic part of the infrastructure, and one that would mark the economic rebirth of the island. However, over the past fifteen years, environmentalists have challenged the project, which would have a devastating ecological impact, in a high-risk area for seismic activity with a high concentration of goods and passenger traffic across the Strait, which would make this large-scale project uneconomical (The Economist 2003; for a more in-depth review, see: Angelini 2011). The bridge has never been built, but the various factions involved in electoral campaigns mention it now and again because the subject arouses strong local feelings of interest and therefore unites voters. The bridge across the Strait is another example of speculative economy, which helps to generate the movement of capital and revenue even though there is no concrete result. Even before the building begins, the design project has managed to open up certain expenditure channels. Suffice it to remember here that at the beginning of the 1980s a company called "Stretto di Messina Spa", was set up, using public capital, was fully staffed, and entirely dedicated

to this project. It is estimated that from 1981 up to the present, the cost of running the company amounted to over €300 million (Morgantini 2016). Between 2000 and 2012, during the Berlusconi governments, the final project was approved, and some of the additional work was tendered out; however, the fall of the government meant activities were suspended, and they still are (De Luca 2016).

2. On this subject, part of the video interview with Signora Concetta in which she tells her story can be seen on YouTube, (https://www.youtube.com/watch?v=BBSh4dYccPs) as part of a research project about the lives of the slum-dwellers carried out by Pasquale Filippone; this video was probably the inspiration behind the television story we refer to. The video opens with Signora Concetta who, despite the heat and the decrepit state of the apartment, appears lively. Sitting on a wooden bench, she welcomes the journalist into her home, chatting to him in Sicilian dialect, and explaining that she has lived there for seventy years (like the majority of the slum-dwellers, therefore, she was not born in her shack). She goes on to explain how she spends each day and how she manages to wash. At one point, she even jokes: "You know where I'll finally get a house, don't you? In the cemetery!". Again, when the journalist asks where she eats, she says: "Oh, I eat here! I take out a little saucepan, I've got plates and crockery, I've got a lot of stuff put away, it's not that I don't have anything!" (This forceful clarification is in distinct contrast to the reference about the one single spoon used to beat back the rats made in the television story). She adds: "for now I'm not really taking care of myself", and she gets upset at the thought of the four cats she had adopted to keep the rats away, and to whom she was very attached and who are now dead. This is the reason for her loneliness.
3. For example, the decades-long criticism that followed the seminal work by Lewis (1966) on how the protestant imaginary has influenced the analysis of poverty and how, implicit in his theory, significant blame has been apportioned to those who live on the margins of society. On this subject, see the observations put forward by Bourgois (2003, p. 64) and Wilson (2012, pp. 182 passim).
4. From the Fascist period, home ownership in Italy was seen as a way to integrate the masses into the order of the State. The regime, therefore, saw the increase in home ownership as a way of extending the base of those who had "interests", in the hope, therefore, that they would aspire to order, to the same order that Fascism produced and guaranteed. Similar ideas are to be found during the transition to the Republic and, again, in the post-war period, when the congressional documents of the Democratic Christian party, the primary Italian governmental party up until the 1990s, adopt this legacy of ideas and electoral management, while incorporating it into a Catholic framework based on family values. (Bortolotti 1978, p. 263).

5. Here, the reference is to that branch of anthropological studies regarding the policies of self-representation, that is, the selection processes and use of public cultural themes aimed at fulfilling the functions and achieving the objectives of a social group, including subaltern ones. If some of the seminal references to this subject have been put forward by Ortner (1984), Marcus and Fisher (1986), Clifford and Marcus (1986), Herzfeld (1987), are the works by Palumbo (2003, 2009) that provide many local examples of this type of subordinate manipulation of reality and are, therefore, more closely related to our analysis.
6. A discursive and aesthetic climax similar to that of "humanitarian reason" as identified by Fassin (2012).
7. Despite some significant differences regarding the generality of the claims, similar dynamics are apparent in the famous study by Holston (2008) about the struggle for the recognition of the right to housing space in the São Paolo favelas.
8. This is a post-democratic government (Crouch 2004), as can be seen by their contempt for parliamentary democracy and the institutional seats that preside over democratic dialogue (Parliament, the Presidency of the Republic, the magistracy and the press). These have been substituted by fake direct democracy, endorsed by the constant personal presence of the leader of the government on personal social networking sites, which he uses to communicate directly with the public. In reality, the public become the passive users of news that is presented without any possibility of contradiction. Deprived of any critical elements that might help them to evaluate the quality or truth of the information received, they can only exercise the right to publish a "like" or a comment. This is a truly limiting form of liberty, in complete contrast with the idea of active participation typical of direct forms of democracy. Furthermore, the impulsive, emphatic, and sensationalist tones used to present the facts transform the public into permanent potential electors, so that they are under the impression of being perpetually involved in an election campaign.
9. There are very close links between the construction industry and the mafia. Regarding the Sicilian case cfr. Sacco (2010).

References

Angelini, A. (Ed.). (2011). *Il mitico Ponte sullo stretto di Messina. Da Lucio Cecilio Metello ai giorni nostri: la storia, la cultura e l'ambiente*. Milano: FrancoAngeli.

Bellicini, L. (1989). In periferia. Temi, percorsi e immagini. *Meridiana, 5*, 93–128.

Bellicini, L. (1990). La costruzione del territorio meridionale. *Meridiana, 10*, 11–44.

Bortolotti, L. (1978). *Storia della politica edilizia in Italia*. Roma: Editori Riuniti.

Bottari, M. (2017, December 12). Tutto fermo al Rione Taormina. *Gazzetta del Sud*, ed. Messina.

Bourgois, P. (2003). *In Search of Respect. Selling Crack in El Barrio*. Cambridge and New York: Cambridge University Press.

Bridges, K. M. (2017). The Deserving Poor, the Undeserving Poor, and Class-Based Affirmative Action. *Emory Law Journal, 66*, 1049–1114.

Brunson, N. (1993). Ideas and Actions: Justification and Hypocrisy as Alternatives to Control. *Accounting, Organizations and Society, 18*(6), 489–506.

Clifford, J., & Marcus, G. (1986). *Writing Cultures. Poetics and Politics of Ethnography*. Berkeley and Los Angeles: University of California Press.

Cremaschi, M. (1990). L'abusivismo meridionale: realtà e rappresentazione. *Meridiana, 9*, 125–154.

Crouch, C. (2004). *Post-Democracy*. Oxford: Polity Press.

De Luca, D. M. (2016, September 28). Breve storia del ponte di Messina. Comincia ai tempi delle guerre puniche e arriva a Matteo Renzi, passando per Carlo Magno e l'ingegner Cane. *Il Post*. Retrieved from https://www.ilpost.it/2016/09/28/storia-ponte-sullo-stretto/.

Fassin, D. (2012). *Humanitarian Reason. A Moral History of the Present*. Berkeley and Los Angeles: University of California Press.

Fera, G., & Ginatempo, N. (1982). *Autocostruzione, marginalità o proposta*. Reggio Calabria: Gangemi.

Fera, G., & Ginatempo, N. (1985). *L'autocostruzione spontanea nel Mezzogiorno*. Milano: FrancoAngeli.

Ferrarotti, F. (1970). *Roma da capitale a periferia*. Bari: Laterza.

Fiumi, C. (2008, April 8). Messina: Cent'anni nelle baracche. *Corriere della Sera*, May 15. Retrieved from https://www.corriere.it/cronache/08_aprile_15/magazine_messina_cento_anni_di_baracche_5a181f44-0af0-11dd-98e1-00144f486ba6.shtml.

Foucault, M. (1978). *The History of Sexuality Volume 1: An Introduction*. New York: Pantheon Books.

Foucault, M. (1988). An Aesthetics of Existence. In M. Foucault (Ed.), *Politics, Philosophy, Culture. Interview and Other Writings 1977–1984*. New York: Routledge.

Foucault, M. (2006). *Psychiatric Power: Lectures at the Collège de France, 1973–1974*. New York: Palgrave Macmillan.

Foucault, M. (2009). *Le Corps utopique, suivi de Les Hétérotopies*. Paris: Nouvelles Éditions Lignes.

Herzfeld, M. (1987). *Anthropology Through the Looking Glass: Critical Ethnography in the Margins of Europe*. Cambridge: Cambridge University Press.

Herzfeld, M. (2005). *Cultural Intimacy*. New York: Routledge.

Holston, J. (2008). *Insurgent Democracy. Disjunctions of Democracy and Modernity in Brazil*. Princeton: Princeton University Press.

Lewis, O. (1966). The Culture of Poverty. *Scientific American, 215*(4), 19–25.

Marcus, G., & Fisher, M. (1986). *Anthropology as a Cultural Critique. Experimental Moment in the Human Sciences.* Chicago: Chicago University Press.

Morgantini, A. (2016, September 29). Ponte sullo Stretto: oltre 300 milioni di euro, ecco quanto è costata sinora la Sdm. *Il Fatto quotidiano.* Retrieved from https://www.ilfattoquotidiano.it/2016/09/29/ponte-sullo-stretto-oltre-300-milioni-di-euro-ecco-quanto-e-costata-sinora-la-sdm/3064294/.

Ortner, S. (1984). Theory in Anthropology Since the Sixties. *Comparative Studies in Society and History, 26*(1), 126–166.

Palumbo, B. (2001). Campo intellettuale, potere e identità tra contesti locali, "pensiero meridiano" e "identità meridionale". *La ricerca Folklorica, 43*, 117–134.

Palumbo, B. (2003). *L'Unesco e il Campanile. Antropologia, politica e beni culturali in Sicilia orientale.* Milano: Meltemi.

Palumbo, B. (2009). *Politiche dell'inquietudine. Passione, feste e poteri in Sicilia.* Firenze: Le Lettere.

Redazione. (2015, May 1). Gli alloggi sono nostri. Così gli abitanti di rione Taormina. Che li occupano. *Normanno.* Quotidiano online di Messina e Provincia. Retrieved from http://normanno.com/cronaca/gli-alloggi-sono-nostri-cosi-gli-abitanti-di-rione-taormina-che-li-occupano/.

Sabot, P. (2012). Linguaggio, società, corpo. Utopie ed eterotopie in Michel Foucault. *Materiali foucaultiani, 1*(1), 17–35.

Sacco, S. (Ed.). (2010). *La mafia in cantiere. L'incidenza della criminalità organizzata nell' economia: una verifica empirica nel settore delle costruzioni.* Palermo: Centro di studi ed iniziative culturali Pio La Torre.

The Economist. (2003, November 20). The Strait of Messina Bridge. A Bridge Too Far. The Questionable Economics of a Grand Italian Construction Project. *The Economist.*

Ward, C. (2002). *Cotters and Squatters. Housing's Hidden History.* Nottingham: Five Leaves Pub.

Ward, C. (2004). The Hidden History of Housing. *History & Policy*, September 1. Retrieved from http://www.historyandpolicy.org/policy-papers/papers/the-hidden-history-of-housing.

Weber, M. (1958). The Three Types of Legitimate Rule. *Berkeley Publications in Society and Institutions, 4*(1), 1–11.

Wilson, W. J. (2012). *The Inner City, the Underclass, and Public Policy.* Chicago: The University of Chicago Press.

CHAPTER 8

Conclusion

8.1 Transcending Events

Disasters are complex events that can be read on two levels. The first one is the local level, entirely centered on the local aspects—the "monographic" layer, we might call it. The other level concerns the connections, both local and global, which relate the single event to dynamics and processes that are apparently external to both the event and the locality—the "extended" layer, so to say. The 1908 Messina earthquake, then, as almost any other disaster of the modern age, provides us with the possibility of seeing the interplay between local and external factors. It can be read as a local event with origins in local characteristics (e.g., faulty architecture), or a story of failed reconstruction due to internal elements (a corrupt political class), and as an "interconnected" disgrace whose outcomes derive from a specific national organization of social, economic, and political relations and exchanges.

Both readings are possible and produce, in fact, different results. The former reading, which is subscribed to not only locally, blames the city and its inhabitants for the outcomes of this (missed) rebirth. This is an interpretation of the event that we have discussed in Chap. 2, and it manifests itself through stories such as the one on the massive emissions of radon gas that could have altered the DNA of the inhabitants of the city. In this reading, the *messinese*, although not entirely responsible for his/her failures, was subjected to an organic change that made it impossible for this citizen to be

D. Farinella, P. Saitta, *The Endless Reconstruction and Modern Disasters*,
https://doi.org/10.1007/978-3-030-19361-4_8

other than the loser he/she is. As we have seen, this interpretation, then, spins around the problems of civility, ethos, and underdevelopment. The latter reading, instead, interprets the earthquake as an event that is situated within a field of forces and has, therefore, political meanings with importance transcending the local level. This reading of the facts, which corresponds clearly to our favorite interpretation, sees the Messina earthquake as a highly metaphoric event that sheds lights on the nature of the political hierarchies and economic interests that are at the basis of the national organization at a given time. Not a local event, but a national one, as Dickie (2008) has aptly noticed. Albeit these are features, in our view, that almost any contemporary disaster presents, and are, thus, quite generalizable.

8.2 Continuity

From what was said above, one could conclude that disasters may be seen as "opportunities". Not in the sense that supporters of folk concepts such as "resilience" assign to this word, nor in the sense that neoliberals following Schumpeter pursue. Perhaps paradoxically, this notion of opportunity entails for us the ideas of inequality and limited access to chances. In fact, disasters are always negative occurrences for most people, with the exception of a few of them. So, when we claim that calamities are opportunities, we focus on the few and suggest that such prospects are provided by the intersection of political and economic openings, which produce an unequal distribution of material and symbolic capital. A distribution process, moreover, that has an impact on the public institutions and is elaborated, justified, and perceived through the rhetoric, instruments, and lenses of public culture. In our view, public culture is the institutional capability of manipulating meanings and producing interpretative frames that are at the basis of the general understanding and reception of institutional activity. Or, in Said's words, "The power to narrate, or to block other narratives from forming and emerging" (Said 1979, p. xiii; on this point, see also: Mulcahy 2017, p. 96). In this latter respect, modern disasters are public events not only for being events that involve entire populations but for their capability of being interpreted, narrated, and managed in the light of the public culture of a given time and, above all, for their scalar logic—that is, for their capability of connecting the local and the central levels, the "top" and the "bottom" of societies. For their capability, as we have said in the previous pages, to reproduce "below" what exists "above". Thus, the production and reproduction of interests as well as of "pariahs" in any corner of a national territory.

In this sense, disasters are the by-product of their epoch. We utilize this word "epoch" rather than other possible synonyms (age, time, etc.) because we would like to stress the importance of a long temporal horizon for the analysis—coherent with the idea of institutional actions as stages of a project or a "scheme" (Scott 1998).[1] An extended period during which the features of a social organization develop, change, and adapt to the contingency but, nonetheless, maintain their traits. The amount of time, in other words, that precedes substantial changes and paves the way to a new epoch—that is, to a new regulatory regime. With this purpose in mind, perhaps artificially, we maintained, following Noto (2008), the distinction between disasters and catastrophes—the former being those undesired events whose effects can be rapidly reabsorbed, and the latter those disgraceful occurrences whose outcomes cannot be reversed and can be fully appreciated only after a long lapse of time. Occurrences that in the act of disfiguring forever a place are, nonetheless, accompanied by logics and interests which are often known, recognizable, and long lasting.

If in these abstract terms the temporal fractures are frankly too vague, by this book, we implicitly suggest that the analysis of a disaster should take place over the course of an extended period of time—over a century, in our case. In the field of disaster studies, moreover, this idea is coherent with Parrinello's work, who, in his study of the Messina's tragedy, has analyzed the thirty years that precede and follow the 1908 earthquake (Parrinello 2015). Following the same logic, we stretched the time taken into consideration for the analysis because we were interested in the notion of continuity and the outcomes of certain choices on the present. That is, we were interested in the full unfolding of the effects of post-disaster interventions and, above all, in the permanence of the logic and materiality of social relations that made this disaster possible and permeates the reconstruction and its effects.

8.3 Integrated Analytical Frameworks

In more concrete terms, what we noticed is that the political logic, structure, and ties that characterized Messina before the crisis survived the earthquake. Likewise, the economic trends that preceded the catastrophe and consisted in their proto-financialization were "augmented" and made suddenly visible. Similarly, the segmentation and segregation that were at the basis of local urbanism prior to the telluric event were reproduced and brought to the extreme. As we have seen, these findings are consistent with what the bulk of the literature has noticed when reflecting on other disasters, whose different outcomes depended on the insertion of the involved cities into more positive frameworks (e.g., Chicago and San Francisco after the Great Fire and the 1906 earthquake, respectively) (Sawislak 1995; Dyl 2017). Frameworks, clearly, whose extension includes at least the national level but whose understanding, in the most extreme cases, should entail an analysis of the international relationships—the case of the "NGO Republic of Haiti" being, for example, only the most visible and famous (Klarreich and Polman 2012). International aid is a good example as it implies matters of diplomacy, nationalistic pride, and symbols. When a tremendous earthquake hit Calabria in 1783, the Bourbons, at least officially, refused international aid (although, in an unofficial way, the Maltese ships had been admitted) (Bottari 2010, p. 48). The royals' intent, clearly, was performative and aimed at showing competency and self-sufficiency before the international community. Likewise, in 1908, the Italian State accepted the help of many states, except, at least initially, Austria. Due to past relationships, this country was suspected of wanting to take advantage of the circumstances and to attack Italy (Mazzoli 2008).

Today, the weight of the international relations affects the recovery of places hit by disaster in other more subtle ways. For example, in terms of extra budgetary resources left available for public interventions. This was the case of Italy after the seismic waves that hit central Italy in 2016 (Emidio di Treviri 2018), when the government was forced to urge the European Union to allow extra fiscal leeway for the country to fund measures aimed at limiting the damage of the earthquake (Politi and Brunsden 2016). Or, the case of Greece before and after the lethal wildfires that hit a vast area close to Athens in 2018. According to the analysts, in fact, 30% of the trucks and the other machinery were off the roads due to technical problems, while the personnel were considerably understaffed due to the missed turnover imposed by the EU. "The debt-stricken country has applied tough measures in return for bailout loans to keep bankruptcy at bay" was the conclusion of a press report (Smith 2018).

Our case study, however, shows also that at the intersection of economy and national relations and cultures (or, supranational cultures, if one thinks of creations such as the European Union with its internal informal hierarchies) lies the "locality", with its historical meanings and functions. Suddenly, the rubble of the 1908 earthquake propelled Messina directly into the context of the "Southern Question", which was at the heart of the national discourse and divisions since the Unification. This was a discourse that depicted the Italian *Mezzogiorno* as a place situated at the borders of civilization (Petrillo 2016). If, on the one hand, this city of ruins became the place to establish a highly segmented society and to exercise a particular kind of disciplinary power based on the informal exchanges between institutions and society, on the other hand, the same city was encircled in more traditional forms within the wider national "Question" and hierarchies. Like most centers of the macro-area to which it belonged (Ferrari Bravo and Serafini 1972; Gribaudi 1990), this formerly influential, rich, and productive city was suddenly transformed in an outlet for the goods produced in the north of the country and, slowly, also in the permanent storage space of a reserve labor army that, through emigration, would help build the country as well as the national industry. Catastrophes, in sum, can be a way to produce new and paradoxical forms of integration within a national scheme.

On the opposite side, the 2012 Emilia Earthquake, which hit one of the most economically relevant regions of the country, produced completely different effects. Given the importance of the region, and the political proximity of the local and national governments (both were at that time Democratic), the State's response was prompt—especially with regard to the economic infrastructure (Pitzalis 2016). Moreover, unlike the "southern earthquakes" from the Unity to the present, this event was accompanied by a public rhetoric entirely centered on the moral temper of this population devoted to hard work, generosity, and resilience.[2] The full integration of Emilia within that national scheme, and its relevance for the national economy, did not allow the disaster to be turned into a permanent catastrophe. Thus, our hypothesis is that the position of a locality within a national scheme is one of the best predictors of the promptness of institutional responses—something, however, that should not be confused with the "quality" of the countermeasures. For example, the response of the New York authorities after September 11 was certainly rapid, in spite of its debatable inclusiveness. As a matter of fact, financial firms were framed as the primary economic victims, and, subsequently, they became the main beneficiaries of redevelopment funds (Gotham and Greenberg 2014, p. ix). Thus, as is perhaps obvious, reconstruction processes reflect also preexisting local differences in class, power, and influence. Something, moreover, that affects also the ways in which victims are identified and assisted and that produces interesting paradoxes such as the one seen in action in New York after the 2001 terrorist attack. The paradox, in other words, that consists in subversion of expectations and in the transformation of the wealthy and powerful into "needy" subjects (and, conversely, in the implicit and factual transformation of the poor into the "undeserving").

8.4 Events Rooted in an Epoch

The Messina earthquake, however, is also a catastrophe of the twentieth century in at least three aspects. Firstly, it has produced different regimes of visibility in relation to classes and territory. Secondly, it replaced the materiality of production with rent. Thirdly, most of its effects were possible in the presence of a welfare state.

With regard to the first point, Brighenti (2010, p. 3) presents visibility as an inherently ambiguous phenomenon dependent upon political arrangements that can be termed "regimes". The term vision entails on the one hand the notion of lived experience and power, and on the other, the uses of this human faculty to coordinate action. Clearly, the author observes, the visible entails more than the visual or the sensorially perceptible. That is, it involves the culture and socially constructed meanings. Yet, the visible is a "field of inscription and projection of social action, a field which can be explored as a territory" (Brighenti 2010, p. 4)—the "territory" being a material way of defining, inscribing, and stabilizing patterns of relation between and within a multiplicity of *socii* (Brighenti 2010, p. 42). In accordance to this conceptualization, our study showed that the Messina reconstruction plan has been a modern attempt to produce a segmented territory that entailed different regimes of visibility for each class. In doing this, the author of the master plan and, later, the Fascist regime followed a rationality that pertained to the modern concept of city. At least since Hausmann's re-writing of Paris in the period 1853–1870, the problem of the planners at the turn of the century was to provide the bourgeoisie with adequate accommodation and, at the same time, to limit the social struggles and questions underlying the process of the reproduction of labor power (Castells 1977, pp. 35–36). These were processes that had been seen in action in many cities and countries where phenomena such as the emergence of new social blocs, the presence of a particular type of "psychological complex" related to the image of the nation and its elites, and the problematic encounters with "modernity" and its rationale were taking place (Holston 1989, p. 5 passim; Rabinow 1995, p. 211 passim; Agnew 2010; Perlman 2010, p. 26 passim; Sundaram 2010, p. 29 passim). If there is an endless literature on these urban processes in the big cities worldwide, the case of Messina shows how even the "province", or the less important cities, were invested with the ambition and the need to re-write their urban space and their regimes of visibility. Subsequently, our study shows how the forms of tolerable proximity between classes and the forms of "invisibilization" of undesirable classes were produced through a certain use of the territory and the erection of a number of slums located far from the city center. A re-writing process of urban space and social relations, however, that in Messina met the same resistances and problems faced by Hausmann and the other great planners in almost any of the abovementioned cities, at the hands of both the bourgeoisie and the subaltern classes (Paccoud 2015; Agnew 2010). An attempt, thus, that in the case of the Sicilian city was only partially achieved, and that, over the course of the decades, resulted in the impossibility of rendering invisible the undesirable. If, at a later stage, some shanties were self-built in the interstices of the very city center—for example, in vacant lots between a palace and another—the expansion process of the city integrated the shantytowns in the new urban landscape, which offered now a mix of visions and styles. A process that resembles that "Third World urbanism" that bespeaks, in Davis's words, of a "treason of the State" and the "reneging upon historic state commitments to relieve poverty and homelessness" (Davis 2007, p. 72).

The second related point concerns the transformation of the local economy and the rise of a class of renters whose wealth is based on real estate. In this respect, we showed how, prior to the earthquake, Messina's economy was going through processes of "early tertiarization" (or a shift from primary industrial production to the service sectors) and "proto-financialization". With the disruption of the productive and trading infrastructure—this latter consisting of a continental and transcontinental network—land, housing, and financial investments became the operational field of the local economic bourgeoisie. Finally, the State, by its bureaucracies and allocations, completed the framework of bourgeoisie occupations and returns. The remaining part was composed mostly of backward economic sectors (the most relevant of which was food processing) and the construction industry—this latter being the main sector of employment of the local skilled and unskilled labor force. This economic structure, which resisted roughly until now, had important consequences on the urban environment and determined the impressive extension of today's city (about sixty kilometers, or thirty-seven miles, from north to south). An extension and an erosion of the territory, which does not correspond to the composition of the population (slightly over 260,000 at the apex of the demographic process in 1981; about 236,000 the current population). High emigration rates, then, were the necessary outcome of a structural condition that leaned toward the marginalization of the city. A structural condition, however, that has been also a prelude to the social polarization of the future years, that is, of our present. Not by chance, the difference and the distance between the protected social groups and their "other", that is, the mass of precarious and menial workers, bespeak of a social division that resembles the analyses of critics like Harvey (2010) or Bauman (2011). In this respect, the dematerialization of Messina's economy on the one hand and its dependence on real estate on the other were processes which have been mistakenly read as signs of backwardness, while, in reality, they were marks of "futurability" (Berardi 2017)—deprived, certainly, of the sense of optimism present in Berardi's original formulation—of this territory.

Yet, Messina's recovery took place in the same years in which a welfare state was developing. The creation of vast neighborhoods filled with barracks and *casette ultrapopolari* was, in Scott's words (Scott 1998), part of a "scheme" that aimed at sheltering people, at preventing the spread of diseases, and at reducing the impact of social protests. Classically, we might claim that in the 1920s these measures projected the city into a biopolitical realm of practices that aimed at pursuing discipline and control of bodies, through both "knowledge" and the creation of a multitude of regulative agencies located all over the levels of social organization (Foucault 1979; Hewitt 1983, p. 69). Moreover, these were primordial forms of welfare that included "little besides poor relief, sanitation and public nuisance" (Titmuss 1963, p. 40; Hewitt 1983, p. 73). Yet, due to the poor character of these early forms of welfare, these neighborhoods resembled the space of the prison or, perhaps, the "camp", as Agamben (1998) put it. With regard to the latter, the philosopher's claim is famous, saying that, "Today it is not the city but rather the camp that is the fundamental biopolitical paradigm of the West" (Agamben 1998, p. 181)—the camp being, in simple words, a deteriorated space that contains a life deprived of rights and subjected to the arbitrary will of the powerful. While we certainly agree with Agamben and find that these neighborhoods sometimes resemble the "camp", at least physically, in terms of quality of life and exposition to an external discretion, we are skeptical regarding the hypothesis that these spaces were truly biopolitical. If it is true that they expressed a logic that was biopolitical and were also settled at the apex of the historical process of circulation of certain ideas on the organization of the margins, indeed, they showed also a pantomimic character. As we said, the relief provided to the population of these shantytowns was minimal, and the life conditions of the residents were, in general, quite bad. Thus, these were not spaces of reform but of containment and confinement and, later, of "exchange". Especially in the most advanced phases of the consolidation process of these spaces, if a "disciplinary" goal led the action of the public authorities, this consisted in the generation of practices that transformed rights into favors and "items" to be exchanged in the political arena—votes for houses, political campaigns for jobs. If under Fascism, the shanties (i.e., the wooden barracks and the *casette ultrapopolari*) were both the sign of State's benevolence and a "concentrationary" universe subjected to soft custody, under the Republic, they became the space of production and reproduction of a vast political clientele. Keeping aside the evidence that at a later stage—hence, far from the acute phase of post-earthquake recovery—shantytowns, that is the camps of barracks, were not part of a welfare apparatus aimed at providing houses, but the institutional sign of exclusion of the dwellers from such measures. And that, at the same time, shanties were the "soil" and the "capital" of a new power that needed the "demos", the people, to climb up the political ladder.

Regardless of the personal "lineage"—whether the individuals and the households that live in the shacks today were born from families that equally had lived in the area or have moved as adults and young couples from a project to a shantytown—these neighborhoods became the permanent infrastructure of a political regime of exchange. From the Christian Democrats of the post-war era to the center-right in the 1990s up to recent days, the shacks were the focus of the feud between a series of political forces whose power depended on the votes of these army of beggars. The political capability of incrementally keeping promises—that is, the capability of providing some, but not everybody, with what had been promised—was the kernel and the art of the complex mechanism of retention of this "clientele". A mechanism, however, that needed resources. It is that dependence on resources that makes this political model a twentieth-century "product" nearing its end. The expiration of resources, the end of welfare as we knew it, and the reduction of political control on the public sector—now subjected to new restrictions and, thus, unable to absorb that unskilled labor force that represented the core of the political clientele—marked the end of a model. Since 2013, the poor neighborhoods showed themselves to be partially free from the control of the classic political parties. Twice, in 2013 and 2018, these areas have voted for "populist" forces that presented themselves as an alternative to the old parties and their methods.[3] If the 2013 vote in support of a leftist populist, Renato Accorinti, resulted in disappointment, in 2018, the elected candidate was a right-wing populist, Cateno De Luca. This latter is a neoliberal populist whose plans seem consistent with his pragmatist ideology. Although his election is too recent and, it is too early to draw a conclusion, in a context of limited resources one of his primary goals is to get rid of the shantytowns. Beyond the civility aspects, the presence of these areas hinders, on the one hand, the "touristification" of the city, and, on the other hand, it prevents new real estate speculations to be made. The hills of the city are overbuilt, and the construction industry is short of land. The acquisition of the lands on which the shantytowns are built would help the city to renew its façade, and it would also help set in motion the local economy, which is, as we have repeatedly said, highly dependent on the construction sector.[4] Without taking into account the positive returns of a similar step on the image of the new mayor, who does not hide his political ambitions. In consideration of the 30,000 vacant houses present in the city, the plan in progress, then, consists of two hypotheses. After having been granted governmental and regional funds destined for the mitigation of the housing question, the first option would be to buy vacant houses from the private market and make them available to the households occupying the shanties at an affordable price (following the criteria that rule the regular public housing allocation system); the second option is to rent houses from the private owners and pay the rent for five years in lieu of the evicted families. The cost of the operation is about €200 million—figures apt to the acquisition of 2000 apartments, comprised between 45 square meters and 110 square meters, at a standard price of €910 for a square meter (Bertè 2018; Ipsale 2018). While both these variants of the project represent a valid contribution to the real estate market and can also be read in terms of "social responsibility" and redistribution, they are in fact consistent with a transnational and short-sighted ideology of our time that aims at getting rid of undesired presences in public space (Smith 1996; Atkinson and Bridge 2005; Semi 2015; Hyra 2017). Should the rent option prevail, it is plausible to ask what, in presence of the actual unemployment rates, these households will do after five years. But, also, the other hypothesis is problematic. It is clear, in fact, that in a few years, the exhaustion of the available funds will have a dramatic impact on the maintenance of the houses, and the city will witness an extension of the dilapidated housing stock, similar to what we have shown in Chap. 5.

However, regardless of the opinion one may have on the project, it is relevant that attention to this topic has heightened in the historical moment when the shantytowns ceased to be electoral basins and chose, so to say, "to revolt" (through their vote). A collective behavior, anyway, that does not follow a sudden realization by the slum-dwellers of the affront suffered for over a century, but, more pragmatically, the sudden, manifest disregard of a centenary informal contract among social actors, and the end of an individualistic resource allocation system. The swan song of that system and its related "social contract" is a gigantic value extraction project that seems to inaugurate the official entry into the neoliberal modernity of this "twentieth-century city". An "entry", however, that is not passive and does not follow, for example, the imposed, universal, and impersonal principles that regulate the labor market or the liberalization process of public services by means of national changes in legislation. Rather, it is an "entry" that stems from a plan conceived locally, which has a collective impact and is, above all, led by that "humanitarian reason" (Fassin 2012) found at the very heart of the contemporary neoliberal processes of constitution of social reality. Processes that are neoliberal because they put the market at the center of their own leading principle, they maximize the instantaneous economic impact of the decisions with little concern for later outcomes, and there is frequently the presence of a "moral" allure that surrounds the projects. While this latter process is nothing new, as it has been witnessed in many other cities (Zukin 1995; Ferrell 2001; Bryman 2004), it is the local rise of this type of politician, with his/her apparatus of techniques, together with the disappearance of the old political cultures and the material resources that supported them, which marks that temporal shift that surrenders Messina to new representations of the modern and desirable.

An implementation model, finally, which is certainly compatible with that notion of "scalability" (Tsing 2015, p. 38) that has shaped our analysis throughout this book (see, in particular, Chap. 2). In Tsing's terms, among other things, scalability requires that the elements of a project are "oblivious to the indeterminacies of encounter". Scalability, in fact, "banishes meaningful diversity, that is, diversity that changes things" (ibid.). The intimate substance of this local example of neoliberal imagination, then, consists of the capability of replicating models by avoiding diversity. It implies, in sum, a peculiar bureaucratization of politics and the transformation of local politicians into "replicators", or, perhaps, technicians, whose main task is understanding local frames not as places of diversity but as fields of implementation of high-profit models that have to generate homogenous social spaces. That is, spaces subtracted from history.

Notes

1. Consistent, above all, with an idea of social process that draws much on Braudel's work. In a 1958 article, recently reproposed in English, Braudel (2009, p. 183) argues that the social sciences "may deal excessively with 'events,' or if you will, they 'presentize' social research, thanks to an empirical sociology that disdains any kind of history and limits itself to short-term data, to on-the-spot surveys".
2. See, among many documents, the video by Anpas, *Appunti di resilienza*, https://www.youtube.com/watch?v=WDdnTv8E1kk.
3. On "populism" in Italy, see: Woods and Wejnert (2014).
4. It is remarkable how the construction industry and the development of real estate and infrastructures are, so to say, disjointed from the market. Houses do not need to be sold, nor rented. Due to the financialization of the sector, and in spite of the bursts, they are an investment in themselves. On this point, see: Guironnet and Halbert (2014).

References

Agamben, G. (1998). *Sovereign Power and Bare Life*. Stanford: Stanford University Press.

Agnew, J. (2010). "Ghosts of Rome": The Haunting of Fascist Efforts at Remaking Rome as Italy's Capital City. *Annali d'italianistica, 28*, 179–198.

Atkinson, R., & Bridge, G. (Eds.). (2005). *Gentrification in Global Context: The New Urban Colonialism?* New York: Routledge.

Bauman, Z. (2011). *Collateral Damage. Social Inequality in a Global Age*. Cambridge: Polity.

Berardi, F. (2017). *Futurability. The Age of Impotence and the Horizon of Possibility*. New York: Verso.

Bertè, D. (2018, August 9). Sbaraccamento, servono 200 milioni. *Gazzetta del Sud*. Retrieved from http://www.gazzettadelsud.it/news/messina/301707/sbaraccamento-servono-200-milioni.html.

Bottari, S. (2010). "L'altro terremoto": Messina, 1783 e dintorni. In A. Baglio & S. Bottari (Eds.), *Messina dalla vigilia del terremoto del 1908 all'avvio della ricostruzione* (pp. 41–56). Messina: Istituto di Studi Storici Gaetano Salvemini.

Braudel, F. (2009). History and the Social Sciences. The Long Durée. *Review, 32*(2), 171–203.

Brighenti, A. M. (2010). *Visibility in Social Theory and Social Research*. New York and London: Palgrave Macmillan.

Bryman, A. (2004). *The Disneyization of Society*. London: Sage.

Castells, M. (1977). *The Urban Question. A Marxist Approach*. London: Edward Arnolds.

Davis, M. (2007). *Planet of Slums*. New York: Verso.
Di Treviri, E. (Ed.). (2018). *Sul fronte del sisma. Un'inchiesta militante sul post-terremoto dell'Appennino centrale (2016–2017)*. Roma: DeriveApprodi.
Dickie, J. (2008). *Una catastrofe patriottica. 1908: Il terremoto di Messina*. Roma and Bari: Laterza.
Dyl, J. L. (2017). *Seismic City. An Environmental History of San Francisco's 1906 Earthquake*. Seattle: University of Washington Press.
Fassin, D. (2012). *Humanitarian Reason. A Moral History of the Present*. Berkeley and Los Angeles: University of California Press.
Ferrari Bravo, L., & Serafini, A. (1972). *Stato e sottosviluppo. Il caso del Mezzogiorno italiano*. Milano: Feltrinelli.
Ferrell, J. (2001). *Tearing Down the Streets: Adventures in Urban Anarchy*. New York: St. Martin's Press.
Foucault, M. (1979). *Discipline and Punish. The Birth of the Prison*. Harmondsworth: Penguin.
Gotham, K. F., & Greenberg, M. (2014). *Crisis and Redevelopment in New York and New Orleans*. New York: Oxford University Press.
Gribaudi, G. (1990). *A Eboli. Il mondo meridionale in cent'anni di trasformazioni*. Venezia: Marsilio.
Guironnet, A., & Halbert, L. (2014). The Financialization of Urban Development Projects: Concepts, Processes, and Implications. *Document de travail du LATTS – Working Paper*, n. 14.
Harvey, D. (2010). *The Enigma of Capital and the Crises of Capitalism*. London: Profile Books.
Hewitt, M. (1983). Biopolitics and Social Policy: Foucault's Account of Welfare. *Theory, Culture and Society, 2*(1), 67–84.
Holston, J. (1989). *The Modernist City. An Anthropological Critique of Brasília*. Chicago: Chicago University Press.
Hyra, D. S. (2017). *Race, Class and Politics in the Cappuccino City*. Chicago: The University of Chicago Press.
Ipsale, M. (2018, August 13). Risanamento, ecco come vendere la propria casa al Comune entro il 25 settembre. *Tempostretto*. Retrieved from https://www.tempostretto.it/news/verso-sbaraccamento-risanamento-come-vendere-comune-propria-casa-entro-25-settembre.html-0.
Klarreich, K., & Polman, L. (2012). The NGO Republic of Haiti. How the International Relief Effort After the 2010 Earthquake Excluded Haitians from Their Own Recovery. *The Nation, 295*(21), 11–17.
Mazzoli, E. (2008). *"Nei giorni di tanta incommensurabile sciagura…". Trieste, l'impero e il terremoto di Messina del 1908*. Milano: Biblion.
Mulcahy, K. V. (2017). *Public Culture, Cultural Identity, Cultural Policy*. New York: Palgrave Macmillan.

Noto, A. G. (2008). *Messina 1908. I disastri e la percezione del terrore nell'evento terremoto*. Soveria Mannelli: Rubbettino.
Paccoud, A. (2015). Planning Law, Power, and Practice: Hausmann in Paris (1853–1870). *Planning Perspectives, 31*(3), 341–361.
Parrinello, G. (2015). *Fault Lines. Earthquakes and Urbanism in Modern Italy*. New York: Berghahn.
Perlman, J. (2010). *Favela. Four Decades of Living on the Edge in Rio de Janeiro*. Oxford and New York: Oxford University Press.
Petrillo, A. (2016). Il destino "federale" del Mezzogiorno nella sociologia. *Cartografie Sociali, 1*(1), 31–83.
Pitzalis, S. (2016). *Politiche del disastro. Poteri e contropoteri nel terremoto emiliano*. Verona: Ombre Corte.
Politi, J., & Brunsden, J. (2016, October 27). Rome Pleads to EU for Leeway on Earthquake Spending. *Financial Times*. Retrieved from https://www.ft.com/content/f51782cc-9c5b-11e6-8324-be63473ce146.
Rabinow, P. (1995). *French Modern. Norms and Forms of Social Environment*. Chicago and London: The University of Chicago Press.
Said, E. (1979). *Orientalism*. New York: Pantheon.
Sawislak, K. (1995). *Smoldering City. Chicagoans and the Great Fire. 1871–1874*. Chicago: The University of Chicago Press.
Scott, J. (1998). *Seeing Like a State. How Certain Scheme to Improve the Human Condition Have Failed*. New Haven and London: Yale University Press.
Semi, G. (2015). *Gentrification: tutte le città come Disneyland?* Bologna: Il Mulino.
Smith, N. (1996). *The New Urban Frontier. Gentrification and the Revanchist City*. London and New York: Routledge.
Smith, E. (2018, July 28). Greek Firefighters Join Public Outcry at "Woeful" Response to Lethal Wildfires. *The Guardian*. Retrieved from https://www.theguardian.com/world/2018/jul/28/greece-wildfires-firefighters-attack-government-response.
Sundaram, R. (2010). *Pirate Modernity. Delhi's Media Urbanism*. London and New York: Routledge.
Titmuss, R. M. (1963). *Essays on the Welfare State*. London: Allen and Unwin.
Tsing, A. L. (2015). *The Mushroom at the End of World. On the Possibility of Life in Capitalist Ruins*. Princeton and Oxford: Princeton University Press.
Woods, D., & Wejnert, B. (Eds.). (2014). *The Many Faces of Populism: Current Perspectives*. Bingeley, UK: Emerald.
Zukin, S. (1995). *The Cultures of Cities*. Cambridge, MA: Blackwell.

Index[1]

[1] Note: Page numbers followed by 'n' refer to notes.

D. Farinella, P. Saitta, *The Endless Reconstruction and Modern Disasters*,
https://doi.org/10.1007/978-3-030-19361-4

The manufacturer's authorised representative in the EU is Springer Nature Customer Service Centre GmbH, Europaplatz 3, 69115 Heidelberg, Germany. If you have any concerns regarding our products, please contact ProductSafety@springernature.com

Printed and bound by CPI Group (UK) Ltd, Croydon, CR0 4YY
30/09/2026
02225652-0003